Computer Architecture and Design Methodologies

Twilight zone of Moore's law is affecting computer architecture design like never before. The strongest impact on computer architecture is perhaps the move from unicore to multicore architectures, represented by commodity architectures like general purpose graphics processing units (gpgpus). Besides that, deep impact of application-specific constraints from emerging embedded applications is presenting designers with new, energy-efficient architectures like heterogeneous multi-core, accelerator-rich System-on-Chip (SoC). These effects together with the security, reliability, thermal and manufacturability challenges of nanoscale technologies are forcing computing platforms to move towards innovative solutions. Finally, the emergence of technologies beyond conventional charge-based computing has led to a series of radical new architectures and design methodologies.

The aim of this book series is to capture these diverse, emerging architectural innovations as well as the corresponding design methodologies. The scope covers the following.

- Heterogeneous multi-core SoC and their design methodology
- Domain-specific architectures and their design methodology
- Novel technology constraints, such as security, fault-tolerance and their impact on architecture design
- Novel technologies, such as resistive memory, and their impact on architecture design
- Extremely parallel architectures

More information about this series at http://www.springer.com/series/15213

Ayantika Chatterjee · Khin Mi Mi Aung

Fully Homomorphic Encryption in Real World Applications

Springer

Ayantika Chatterjee
Indian Institute of Technology Kharagpur
Kharagpur, India

Khin Mi Mi Aung
Institute for Infocomm Research
A*STAR
Singapore, Singapore

ISSN 2367-3478   ISSN 2367-3486   (electronic)
Computer Architecture and Design Methodologies
ISBN 978-981-13-6392-4   ISBN 978-981-13-6393-1   (eBook)
https://doi.org/10.1007/978-981-13-6393-1

Library of Congress Control Number: 2019930570

This Springer imprint is published by the registered company Springer Nature Singapore Pte Ltd.
The registered company address is: 152 Beach Road, #21-01/04 Gateway East, Singapore 189721, Singapore

Contents

Chapter 1
Introduction

Going to the cloud, has always been the dream of man.

This is the era of communication and data sharing, which requires secure and confidential transactions. Cryptography is the science that concerns the maintenance and analysis of security, which allows authorized access to information (Stinson 2005). Cloud computing is one of the new applications, where cryptography is expected to unveil its power to provide desired security solutions. In particular, fully homomorphic encryption (FHE) is one of the main pillars to exploit the complete advantages of cloud computing.

Cloud computing is a promising innovation to provide public platforms for storing large amount of data. In spite of enjoying the convenience of shared resources, security issues arising due to storing critical data in cloud is a potential threat (Rass and Slamanig 2013). Confidentiality of information in a public domain like cloud is a major concern, which is usually provided by applying suitable encryption (Xiong et al. 2014). However, for every single processing on encrypted data, the data should be downloaded and decrypted in the client side and after processing it should be further encrypted and uploaded onto the cloud. This obvious need for repeated decryption-encryptions increases the processing complexity and out-weighs the advantage of using cloud resources. Since the computation is done in the client end, the objective of utilizing large processing power at the cloud-end is defeated and the information, albeit ciphered, is continuously exposed to the adversary.

ARM TrustZone or Intel SGX like trusted platforms provide an isolated secure execution environment (Ekberg et al. 2013). However, all these platforms are based on some trusted applications and designs can defend against only a subset of the possible attacks. Homomorphic encryption scheme provides a mechanism to perform operations on encrypted data without decrypting it. Moreover, FHE allows arbitrary operations on encrypted data, hence it is considered as the holy grail of cryptography.

© Springer Nature Singapore Pte Ltd. 2019
A. Chatterjee and K. M. M. Aung, *Fully Homomorphic Encryption
in Real World Applications*, Computer Architecture and Design Methodologies,
https://doi.org/10.1007/978-981-13-6393-1_1

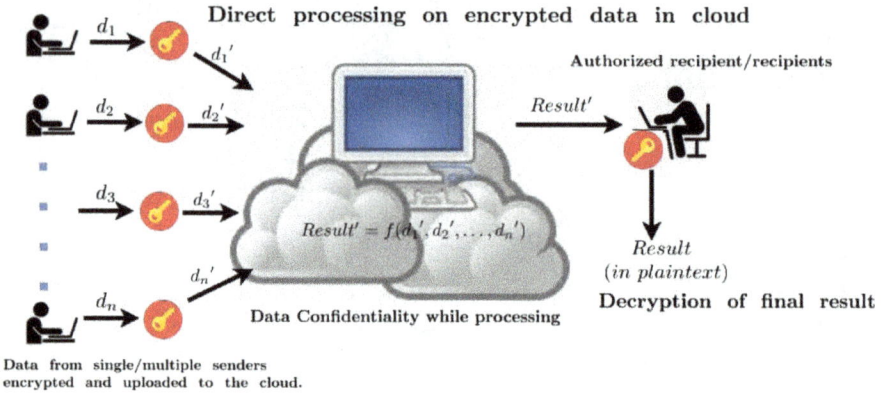

Direct processing on encrypted data in cloud

Fig. 1.1 FHE processing on cloud

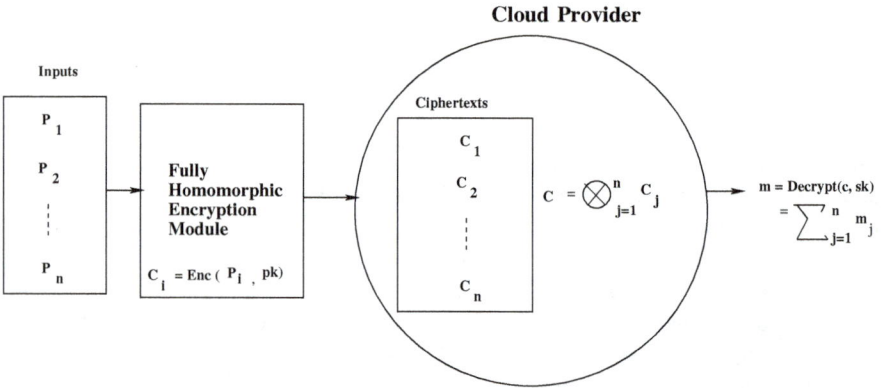

Fig. 1.2 FHE processing on cloud data (Rass and Slamanig 2013)

Thus, the main motivation of our book is to design efficient algorithms for databases where data is already FHE encrypted, so that any arbitrary operations are possible on encrypted data as and when required in the cloud as shown in Fig. 1.1. While group homomorphism provides capability to execute some restricted computations over encrypted data. As shown in Fig. 1.1, user/users can upload data to cloud encrypted with their respective keys and encrypted computation can take place (under single homomorphic key or by secure multiparty computation) in cloud with the power of homomorphic encryption. The encrypted result can be downloaded and decrypted in plaintext as and when required.

Figure 1.2 as depicted in Rass and Slamanig (2013) illustrates the advantage of homomorphic encryption in cloud. The diagram shows that plaintexts $p_1, p_2 \ldots p_n$ are encrypted with FHE and the ciphertexts $c_1, c_2 \ldots c_n$ are uploaded in cloud, where $c_i = Enc(p_i, pk)$ and pk is the public key. Further, direct processing on ciphertexts are done in the cloud and no separate decryption-encryptions are required. The final ciphertext is denoted as $c = \otimes_{j=1}^{n} c_j$, where $\otimes$ is the corresponding operation in the ciphertext space for an operation, $\sum$ in the plaintext space. Thus, upon decryption the resultant ciphertext c should correspond to the plaintext m, which is a result of applying a corresponding operation $\sum$ on the plaintext, thus $m = \sum_{j=1}^{n} m_j$. Naturally, the use of homomorphic encryptions has lots of promises for computations in a public cloud, providing both security and privacy to users.

The first possible construction of performing arbitrary manipulations like addition, multiplication on encrypted data without the knowledge of secret key was mentioned in 2010 by Gentry (2010) and it is termed as fully homomorphic encryption (FHE). Thus generalizing the above discussion from specific operators to any arbitrary function f, we can state the objective of FHE as follows.

Consider the messages $p_1, \ldots, p_t$, which are encrypted to the ciphertexts $c_1, \ldots, c_t$ with the FHE scheme under some key. For any efficiently computable function f, the FHE scheme allows anyone to efficiently compute a ciphertext that decrypts to $f(p_1, \ldots, p_t)$ under the secret key. In spite of its promise, there are several limitations of FHE, with performance being a severe bottleneck. Though the definition of FHE provides the methods to perform arbitrary computations, realizing algorithms to operate on encrypted data using FHE seems to pose several challenges. This inspires to develop suitable synthesis techniques to handle algorithms which operate on FHE data and find methodologies to improve their performances. In the next section, few potential applications for homomorphic encryption HE will be mentioned to show the breath of importance for designing algorithms in different domains supported by underlying FHE encryption.

1.1 Homomorphic Encryption in Real Applications: Few Case Studies

In this section, we mention few real life applications (Archer et al. 2017) where usage of computation in encrypted domain is of utmost importance for sharing data with privacy:

- **HE in National security**: Power of encrypted computation can enhance national security in different ways. Suppose, in case of smart grid network each node from individual generator/building/microgrid need to be monitored by larger smart grid/municipality/government. Since, the analysis performed on the data coming from such microgrids is used to further control the overall distribution of power, it is very important to confirm that such data should not be tampered while transmitted or outsourced to the monitoring platform (like cloud) for analysis and other computational work. In this case, homomorphically encrypted measurements from each node in the grid can be sent to monitoring platform to prevent potential attacks. Another practical application of HE is in case of nationalized database design, where security is an important issue. Such huge dataset can be the prime focus of adversaries and hence need to be stored in encrypted form to maintain confidentiality against different cyber threats. Further, use of HE can provide the power of encrypted domain computation to monitor and analyze such data for nationalized decision making.

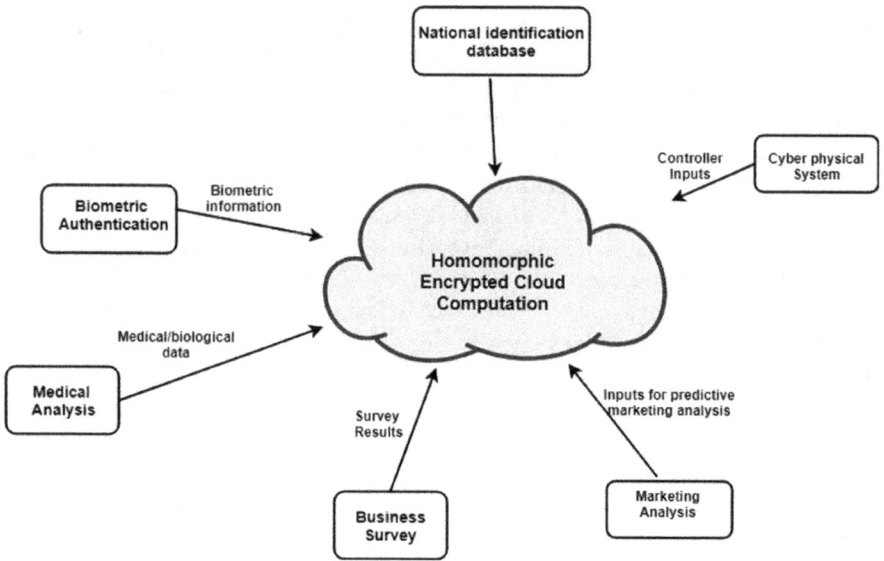

Fig. 1.3 FHE Applications

- **HE in Predictive Analysis**: Predictive analytics is an advanced tool nowadays to forecast behavior and future trends using new and historical data. This is very useful to predict drop out risk for an educational institute, to make crime related predictions for police departments, to estimate different future trends for business and welfare organizations and many more. Data confidentiality risk is associated while creating single repository for sensitive information on which such predictive analysis need to be performed. Hence, adopting the use of HE can provide confidentiality as well as may support to run the algorithms directly on encrypted data (Fig. 1.3).
- **HE in Medical**: Medical information is highly sensitive, yet need to be shared for different purposes. Hence, healthcare industry is one of the most prominent domain for HE application. Every year larger amount of data-breaches exposing medical records lead to cost of huge compensations for medical organizations. Cyber insurance is a way to provide some protection against such damages. However, that also provides minimum policy values in exchange of substantial premiums. Medical data also need to be shared between health insurance companies for policy details determination, hospitals and medical organizations to extend medical research scope. Again, sharing data with privacy is an important requirement for enhancing genomic study. Human DNA and RNA sequences are biometric identifiers like a fingerprint and they can reveal critical information such as disease risk or socially for the identification of family of diseases, such as the presence of an Alzheimer's allele or the discovery of non-paternity. Present strategies for genomics data pro-

tection are having high overhead on researchers, hence homomorphic encryption
can be highly suitable for genomics data sharing.

- **HE in securing systems**: Modern cyber physical systems (CPS) consist of sensors, actuators and controllers where most of the controllers are cloud based or networked. Different reports show how hackers can remotely take charge of the control systems and manipulate the basic functionality of the CPS according to their own wish. Recently, several researchers suggested use of HE for protecting control systems by encrypting sensor data with HE. Controllers will be able to process this data directly in the encrypted domain enhancing overall security.
- **HE in Marketing and Business**: One recent survey by Microsoft shows, 11% of a company's IT budget is spent on developing new applications, however the rest goes to maintenance and infrastructure. That is the main reason why public services and consumer marketing linked with Cloud Computing grew from 9 billion to 40 billion just in five years. Due to fast and flexible deployment cloud is a promising solution but security is main concern. This security issue becomes critical when analysis/ analytics to be performed on confidential business data in order to make business predictions. HE is a promising solution in this case to maintain data security by performing cloud analytics in encrypted domain.

All these as summarized in Fig. 1.3, require different types of secure computation in encrypted domain. Some of them are database handling operations, some are calculating descriptive statistics, some may require linear predictive model, a tree based model, or a neural network where some may have real time implementation challenges. Theoretically, with the support of FHE one can evaluate any function on encrypted data without divulging unencrypted data to the adversary. However, it is stated in Rass and Slamanig (2013) that "*It must be emphasized that homomorphy is a theoretical achievement that merely lets us arithmetically add and multiply plaintexts encapsulated inside a ciphertext. In theory, this allows the execution of any algorithm complex manipulations like text replacements or similar, but putting this to practice requires the design (compilation) of a specific circuit representation for the algorithm at hand. This may be a nontrivial task.*" The main motivation of this book stems out from the fact that for developing suitable tools to execute algorithms operating on FHE data on general purpose computers, one also needs to architect suitable translations of algorithms operating on unencrypted data to those which operate on FHE encrypted data. This requirement derives from the fact that the FHE schemes are by design *circuit-based* and are not amenable to a non-circuit computation (Xiao et al. 2012). However, classic algorithms are mostly non-circuit based, implying that they are not described in terms of logical gate level operators, like AND-OR gates and multiplexers. Hence one should develop a mechanism by which arbitrary algorithms (which are sequences of instructions to solve a problem) can be executed over FHE encrypted data directly. The ability of performing arbitrary algorithms on encrypted data gives a technology for supporting the much needed privacy-enhancing encrypted operations on cloud. Hence, for proper adoption of FHE for cloud and other potential applications, it is imperative to consider the intricacies of executing the algorithms operating on encrypted data on general purpose computers, which are not equipped with encrypted ALU or does not have a memory where the addressing is encrypted. In order to achieve such an objective we need to consider the challenges of executing such programs on common processors, develop suitable data-structures to handle encrypted data, define efficiency criteria for encrypted algorithms, translate conventional algorithm paradigms to their encrypted versions, and also develop techniques for providing suitable execution criteria of the encrypted program on the host machines which execute the algorithms without knowing the secret key needed to perform decryption. Since FHE supports arbitrary operations on encrypted data, it is the first choice as an underlying encryption scheme to support implementation of different algorithms on encrypted data residing on cloud.

1.2   Summary of This Book

Significant research is being performed to make the FHE based schemes more efficient and practical. Existing FHE libraries support bit-wise homomorphy by bit-wise encrypted addition and multiplication. The basic objective of this book is little different, where four interrelated research issues have been addressed to strive for practical applicability of FHE in general algorithms.

1. The first technique of this book discusses how to handle the age old problem of sorting on FHE encrypted data.It further explores how to develop comparison based sorting using FHE swap circuit, which is comprised of FHE primitives (like adder, multiplier). This discussion also highlights that the greater-lesser relation between inputs of the swap operation remains encrypted and secret to an adversary. With such observations, we relate the problem of sorting to security against chosen plaintext attack (CPA) and answer a pertinent question why partition sort is not a better choice over comparison sort while working with encrypted data. Further, a two-stage sorting approach, named as *Lazy Sort* will be discussed to show how this FHE specific sorting technique leads to performance improvement.
2. Next topic of this book will be on handling any arbitrary algorithms on FHE data and highlight related challenges. Starting from realizing the variables in encrypted domain, translation of Arithmetic, Relational, Logical, Bitwise and Assignment operators as well as encrypted loop handling are the major focus. Further, it is highlighted that algorithms on encrypted data bring forward many new challenges in terms of stack handling, recursion implementation while handling encrypted conditions on existing underlying unencrypted processors, and propose viable solutions to these problems.
3. In the subsequent chapter, we shall discuss why FHE is important for cloud database security and fundamentals of FHE encrypted database related operator design.

4. While working with encrypted algorithm translation, detecting encrypted program termination is a major challenge in case of working with processors where the memory addressing is not encrypted. In the final part of this book, we shall focus on explaining the importance and limitations of encrypted processors for supporting operations on encrypted data. In this context, CPA insecurity of existing partial homomorphic encryption based processor and functional limitations of multi-instruction processor motivate us to discuss about FHE encrypted reduced instruction set processor.

References

Archer D, Chen L, Cheon JH, Gilad-Bachrach R, Hallman RA, Huang Z, Jiang X, Kumaresan R, Malin BA, Sofia H, Song Y, Wang S (2017) Applications of homomorphic encryption

Ekberg JE, Kostiainen K, Asokan N (2013) Trusted execution environments on mobile devices. In: Proceedings of the 2013 ACM SIGSAC conference on computer & communications security, CCS'13. ACM, New York, pp 1497–1498

Gentry C (2010) Computing arbitrary functions of encrypted data. Commun ACM 53(3):97–105

Rass S, Slamanig D (2013) Cryptography for security and privacy in cloud computing. Artech House Inc, Norwood

Stinson DR (2005) Cryptography: theory and practice, Third Edition edn. Chapman & Hall/CRC, Boca Raton (Discrete Mathematics and Its Applications)

Xiao L, Bastani O, Yen IL (2012) An efficient homomorphic encryption protocol for multi-user systems. IACR Cryptology ePrint Archive

Xiong J, Liu X, Yao Z, Ma J, Li Q, Geng K, Chen PS (2014) A secure data self-destructing scheme in cloud computing. IEEE Trans Cloud Comput 2(4):448–458

Chapter 2
Literature Survey

Fully homomorphic encryption (FHE) scheme enables computation of arbitrary functions on encrypted data, hence considered as "holy grail" of modern cryptography. This chapter presents the relevance of FHE in present day cloud computing, a brief history of different homomorphic encryption schemes and formally defines fully homomorphic encryption along with basic idea behind Gentry's construction. Gradually, few recent improvements will be described in this area those are proposed following Gentry's scheme but with simpler constructions and better efficiency.

2.1 FHE in Cloud Computing

Cloud computing provides a new way to shift computing and storage capabilities to external service providers with on-demand provisioning of software and hardware at any level of granularity. In recent years, privacy of cloud data and the computations thereon has emerged as important research area in the domain of cloud computing and applied cryptographic research (Rass and Slamanig 2013). Comparatively cheap and easy way of providing security to cloud data is to rely on trusted server-side hardware which is established by Goldreich and Ostrovsky (1996). However, use of secure hardware (Sahai 2008) restricts the outsourcing computation capability of cloud server due to the limited availability and resource constraint. In literature of cloud computing, garbled circuits (Yao 1982; Kolesnikov et al. 2009; Goldwasser et al. 2013) show how to realize secure two-party computation. The main limitation of garbled circuits (GC) is that they need to be rebuilt by the client for different inputs. However, Gennaro et al. Gennaro et al. (2010) show how GC can be reused safely for different inputs with use of FHE. Further, and concepts of twin cloud (Bugiel et al. 2011) and token based cloud computing (reza Sadeghi et al. 2010) are other approaches of cloud privacy with secure hardware but these methods are hard

© Springer Nature Singapore Pte Ltd. 2019
A. Chatterjee and K. M. M. Aung, *Fully Homomorphic Encryption
in Real World Applications*, Computer Architecture and Design Methodologies,
https://doi.org/10.1007/978-981-13-6393-1_2

to parallelize. For this reason, researchers are investigating approaches to enhance privacy of cloud data in different directions and homomorphic encryption is one of them. Hence, study of FHE in real-life algorithms is of great importance.

In the next section, we elaborate the mathematical background and few related works of FHE starting from basics of homomorphic encryption schemes.

2.2 Mathematical Background

Definition 2.1 A *group* denoted by $\{\mathbb{G}, \cdot\}$, is a set of elements $\mathbb{G}$ with a binary operation '$\cdot$', such that for each ordered pair (a, b) of elements in $\mathbb{G}$, the following axioms are obeyed (Stallings 2005; Fraleigh (2002)):

1. *Closure*: If $a, b \in \mathbb{G}$, then $a \cdot b \in \mathbb{G}$.
2. *Associative*: $a \cdot (b \cdot c) = (a \cdot b) \cdot c, \forall a, b, c \in \mathbb{G}$.
3. *Identity element*: There is a unique element $e \in \mathbb{G}$ such that $a \cdot e = e \cdot a = a$, $\forall a \in \mathbb{G}$.
4. *Inverse element*: For each $a \in \mathbb{G}$, there is an element $a' \in \mathbb{G}$ such that $a \cdot a' = a' \cdot a = e$.

Definition 2.2 If $\mathbb{G}$ and $\mathbb{H}$ are groups, a *homomorphism* from $\mathbb{G}$ to $\mathbb{H}$ is a function $f : \mathbb{G} \rightarrow \mathbb{H}$ such that $f(g_1 \odot g_2) = f(g_1) \otimes f(g_2)$ for any elements $g_1, g_2 \in \mathbb{G}$. $\odot$ denotes the operation in $\mathbb{G}$ and $\otimes$ denotes an operation in $\mathbb{H}$ (Rass and Slamanig 2013).

Definition 2.3 *Group homomorphic encryption schemes* Rass and Slamanig (2013) are public key encryption schemes that compute an operation on ciphertexts equivalent to some binary operation on the corresponding plaintexts. If $M \in (\mathbb{H}, \odot)$ is the set of plaintexts to be encrypted under a public key pk into a ciphertext space $(\mathbb{C}, \otimes)$, it holds that $\forall m_1, m_2 \in M$:

$Encrypt(m_1 \odot m_2; pk) = Encrypt(m_1; pk) \otimes Encrypt(m_2; pk) = c_1 \otimes c_2$.

Further, for any pairs of ciphertexts $c_1 = Encrypt(m_1; pk)$, $c_2 = Encrypt(m_2; pk)$, secret key sk and public key pk, $Decrypt(c_1 \otimes c_2; sk) = m_1 \odot m_2$.

Thus repeatedly performing the operation $\otimes$ on ciphertexts results another valid ciphertext, which is equivalent to performing operations on corresponding plaintexts. In literature, the concept of homomorphism is not new. In 1978, the concept of homomorphism was first realized in *RSA* by cryptosystem (Rivest et al. 1978a).

Homomorphism in RSA

In RSA cryptosystem, if the *RSA* public key is modulus m and exponent e, then the encryption of a message x is given by:

$$\mathscr{E}(x) = x^e \mod m \tag{2.1}$$

Considering two messages x_1 and x_2, the homomorphic property is exhibited as:

$$\mathcal{E}(x_1) \cdot \mathcal{E}(x_2) = x_1^e x_2^e \bmod m = (x_1 x_2)^e \bmod m = \mathcal{E}(x_1 \cdot x_2). \qquad (2.2)$$

Since, $\mathcal{E}(x_1 \cdot x_2)$ is a valid ciphertext under RSA, multiplicatively homomorphic property is present in this cryptosystem.

Homomorphism in Paillier Cryptosystem

In *Paillier encryption scheme* (Paillier 1999), considering public encryption key (n, g), a random $r \in \mathbb{Z}^*$, the ciphertext of a message $m \in \mathbb{Z}$ can be computed as:

$$Encrypt(m; pk) = g^m.r^n \ (\bmod n^2) \qquad (2.3)$$

Considering, two ciphertexts $c_1 = Encrypt(m_1; pk) = g^{m_1}.r_1^n \bmod n^2$ and $c_2 = Encrypt(m_2; pk) = g^{m_2}.r_2^n \bmod n^2$, $c_1.c_2$ can be computed as:

$$c_1.c_2 = g^{m_1}.r_1^n.g^{m_2}.r_2^n = g^{m_1+m_2}.(r_1.r_2)^n \bmod n^2 = c_3 \qquad (2.4)$$

where c_3 is a valid ciphertext under Paillier encryption scheme and thus this cryptosystem exhibits additive homomorphism. Among some other encryption schemes, *Goldwasser–Micali encryption* Goldwasser and Micali (1982) proposed in 1982 and *ElGamal encryption scheme* Gamal (1985) proposed in 1984 shows additive and multiplicative homomorphism respectively. The mentioned homomorphic schemes support either additive or multiplicative homomorphism, hence termed as *partial homomorphic encryption* (PHE). ElGamal encryption scheme requires an added discussion due to its specific features in terms of the homomorphic property.

Homomorphism in ElGamal Encryption

ElGamal encryption is used with prime modulus, $p = 2q + 1$ and can be described by:

$$E : G_q \rightarrow G_q \times G_q \qquad (2.5)$$

where G_q is a subgroup order q of $(Z_p)^*$, the multiplicative group of integers modulo p. Encryption function $E(m)$ for ElGamal can be specified as:

$$E(m) = (g^r, m * h^r) \qquad (2.6)$$

where,

- x: secret key
- h: public key
- g: generator
- m: message
- r: randomness

Now, we shall see how ElGamal is multiplicatively homomorphic: $E : (G_q, *) \rightarrow (Gq \times Gq, *)$.

$$E(m_1) * E(m_2) = (g^{r_1}, m_1 * h^{r_1})(g^{r_2}, m_2 * h^{r_2})$$
$$= (g^{r_1+r_2}, m_1 * m_2 * h^{r_1+r_2})$$
$$= E(m_1 * m_2) \tag{2.7}$$

Equation 2.8 ElGamal encryption is in fact homomorphic with respect to multiplication. The multiplicative homomorphic property of standard ElGamal is used in e-voting protocal as implemented in UniCrypt, a Mathematical Crypto-Library (UniCrypt 2017). In this voting protocol, a re-encryption mixnet is employed, where each mixing authority permutes and re-encrypts votes handling them as input to the next authority with the help of multiplicative homomorphic property of ElGamal.

Another unique property of ElGamal encryption is that with a small change it can act as additive homomorphic scheme instead of multiplication. In this case, encryption is defined as $E : Z_q \to G_q \times G_q$ and $E(m) = (g^r, g^m \Delta h^r)$. Now, the additive homomorphic property is retained in the following way:

$$E(m1) * E(m2) = (g^{r_1}, g^{m_1} * h^{r_1})(g^{r_2}, g^{m_2} * h^{r_2})$$
$$= (g^{r_1+r_2}, g^{m_1} g^{m_2} * h^{r_1+r_2})$$
$$= E(m_1 + m_2) \tag{2.8}$$

However, in this process g^m has to be decoded to yield m after decryption.

Till 2009, some already proposed schemes supported a bounded amount of additions and multiplications simultaneously, contrarily to the aforementioned schemes (Martins et al. 2018). Among them, one notable work is of Boneh et al. published in 2005. It is based on the El Gamal cryptosystem but instantiated over Elliptic Curves (ECs). An EC is a mathematical variety with an algebraic group, with a binary operation named point addition. This operation can be applied repeatedly to perform point multiplication which further computes exponentation in El Gamal. Using Boneh et al. (2005), the addition of the ciphertexts resulted in the addition of the underlying plaintexts. Then, subsequent homomorphic multiplication is possible using bilinear pairings to convert points from ECs to Finite Fields (FFs). No bilinear pairings are known that can be applied to FFs, hence no further multiplications are possible. Further, *Somewhat homomorphic encryption* (SHE) Rass and Slamanig (2013) is an extension of PHE which allows an arbitrary number of one operation but only a bounded number of second operation on the ciphertext.

2.2.1 Somewhat Homomorphic Encryption

The first known efficient SHE scheme is proposed as Boneh–Goh–Nissim (BGN) scheme (Boneh et al. 2005), which is capable of handling arbitrary additions and single multiplications. Other approaches of SHE schemes are proposed in the work Melchor et al. (2008) and also in Sander et al. (1999). Here we start with a general encryption scheme and explain why the mentioned scheme is considered to be somewhat homomorphic (Gentry 2009b).

Considering an encryption scheme with odd number p as shared secret key:

- To encrypt a bit m, a random large q and small r are chosen and output ciphertext is computed as $c = pq + 2r + m$. Ciphertext is close to a multiple of p and m = LSB of distance to nearest multiple of p.
- The decryption is performed as : $m = (c \bmod p) \bmod 2$

Considering two ciphertexts $c_1 = q_1 p + 2r_1 + m_1$ and $c_2 = q_2 p + 2r_2 + m_2$ under this encryption scheme, the addition and multiplication operations are defined with the following equations. Addition of the ciphertexts can be defined as:

$$c_1 + c_2 = (q_1 + q_2)p + \underbrace{2(r_1 + r_2) + (m_1 + m_2)}_{\text{Distance to nearest multiple of } p} \qquad (2.9)$$

$$c_1 + c_2 \bmod p = \underbrace{2(r_1 + r_2)}_{\text{error-term}} + (m_1 + m_2) \qquad (2.10)$$

Multiplication of the ciphertexts can be defined as:

$$c_1 * c_2 = (c_1 q_2 + q_1 c_2 - q_1 q_2)p + 2(2r_1 r_2 + r_1 m_2 + m_1 r_2) + m_1 * m_2 \qquad (2.11)$$

$$c_1 * c_2 \bmod p = \underbrace{2(2r_1 r_2 + \ldots)}_{\text{error-term}} + m_1 * m_2 \qquad (2.12)$$

Considering such type of encryption scheme with a noise parameter attached to each ciphertext Homomorphic property retains till the error-term is within a certain limit. However, the error term (or noise) grows faster for multiplication compared to addition. Hence, the scheme is somewhat homomorphic, since it is capable of handling of large number of additions and few multiplications. To be specific, each homomorphic addition adds the underlying noises, and each multiplication multiplies them. Hence, noise growth limits the amount of operations that can be accomplished. This limit could be evaded when the noise begins to approach the critical threshold, the data could be decrypted and reciphered. In this way, resetting the noise to its original low level would require access to the secret key. That diminishes the purpose of FHE. The main magic of FHE is bootstrapping, a technique wherein a cryptosystem evaluates its own decryption circuit homomorphically by integrating in the public key with encrypted version of secret key and this is the way to reset the noise level homomorphically (shown in Fig. 2.1).

Technically Gentry's initial FHE schemes are largely in this area, where limitations of SHE raises a pertinent question if it is possible to define a scheme ε with an efficient algorithm $Evaluate_\varepsilon$ that, for any valid public key pk, any circuit C (not just a circuit consisting of either additive or multiplicative gates) , and any ciphertexts $c_i \leftarrow Encrypt_\varepsilon(m_i; pk)$, outputs $c \leftarrow Evaluate_\varepsilon(C, c_i, pk)$, which is a valid encryption of $C(m_1, ..., m_t$ under pk. Considering these features, Fully homomorphic encryption is defined as follows:

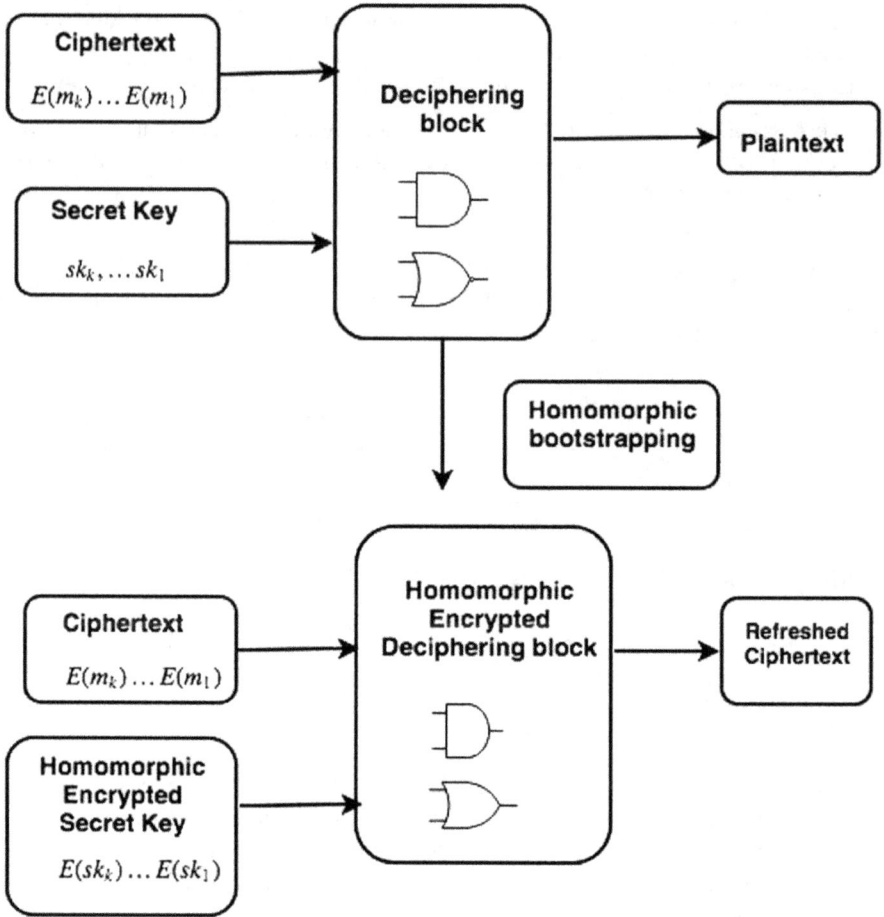

Fig. 2.1 FHE Bootstrapping Martins et al. (2018)

Fully Homomorphic Encryption

Fully homomorphic encryption (FHE) is the most sophisticated homomorphic encryption scheme which allows to evaluate arbitrary functions on ciphertexts. Any public key homomorphic encryption scheme can be defined by the following algorithms (Zhang 2014):

- $KeyGen_\varepsilon(\lambda)$: This probabilistic algorithm takes a security parameter (λ) and produces and outputs a public key pk and a secret key sk.
- $Encrypt_\varepsilon(m, pk)$: This probabilistic algorithm takes a messages $(m \in 0, 1)$, pk and outputs a ciphertext $c = Encrypt_\varepsilon(m; pk)$.
- $Decrypt_\varepsilon(c, sk)$: This deterministic algorithm takes a ciphertext c, secret key sk and outputs message $m = Decrypt_\varepsilon(c; sk)$.

- $Evaluate_\varepsilon(f, c_1, \ldots c_t; pk)$: This algorithm takes pk, n ciphertexts $c_1, c_2 \ldots, c_n$ and a permitted circuit C^n and outputs $C^n(c_1, c_2 \ldots, c_n)$. Considering a set of ciphertexts $\{c_i\}$ whose corresponding messages are $\{m_i\}$, and a circuit C, this algorithm outputs another ciphertext c. This evaluation is correct if the following holds:

$$Decrypt_\varepsilon(Evaluate_\varepsilon(C, c_i, pk), sk) = C(m_1, \ldots, m_t). \qquad (2.13)$$

Definition 2.4 The scheme $\zeta = (KeyGen, Encrypt, Decrypt, Evaluate)$ is homomorphic for a class C of circuits if it is correct according to Eq. 2.13 for all circuits $C' \in C$. ζ is fully homomorphic if it is correct for all Boolean circuits. Further, ζ is compact, if for any circuit $C' \in C$ with a number of inputs polynomial in λ, the size of ciphertexts output by $Evaluate$ is bounded by a fixed value which is polynomial in λ.

Gentry's Framework to Construct FHE

Gentry has shown the first concrete construction of FHE Gentry (2009a, b). These are the steps to construct FHE starting from a SHE scheme Rass and Slamanig (2013):

- Considering a SHE scheme which is capable of evaluating l-variate polynomials homomorphically, where l is considered to be security parameter.
- As shown in Gentry (2009b), bootstrapping technique requires SHE scheme with the capability of evaluating its own decryption function (bootstrappable encryption scheme) plus an additional operation. Thus, SHE scheme is transformed to leveled FHE. This process of evaluating own decryption circuit also requires *squashing the decryption circuit*, which transforms one scheme into another with same homomorphic capacity, but a decryption circuit that is simple enough to allow bootstrapping.
- Figure 2.1 illustrates the concept where ciphertext $E(m_k) \ldots E(m_1)$ and secret key $sk_k, \ldots sk_1$ can perform traditional deciphering to generate plaintext. In bootstrapping, encryption of this secret key $E(sk_k, \ldots, E(sk_1)$ can homomorphically decrypt to generate another ciphertext where plaintext is encrypted with homomorphic encryption key. This notion is called *circular security*. By the notion of *circular security*, the secret key of the bootstrappable encryption scheme is encrypted under the scheme's own public key, and that helps to yield FHE scheme.

Before uncovering the other features of FHE, it is important to mention another important term in FHE literature that is *Recrypt*. This term will be referenced in onward discussions.

Recrypt and Bootstrapping

A key aspect of fully homomorphic scheme from the introduction of Gentry's scheme Gentry (2009b) is the ciphertext refreshing technique, named Recrypt operation. Random noise component grows in size as the ciphertext is processed to homomorphically evaluate a function on its plaintext. Once the noise size in the ciphertext exceeds a certain threshold, the ciphertext can no longer be decrypted correctly. This restricts the number of operations in homomorphic domain to be performed. Recrypt allows refreshing a ciphertext. Given a ciphertext c_i for some plaintext m_i, computing new ciphertext c_n for m is a major objective maintaining the size of the noise in c_n is smaller than the size of the noise in c_i. Repeated refreshing of ciphertexts (e.g., after computing each level of homomorphic operation), one can then evaluate arbitrarily large circuits. Recrypt operation can be realized by evaluating the decryption circuit of the encryption scheme homomorphically. This homomorphic computation of the decryption circuit is also possible without refreshing any ciphertext, and the scheme is called bootstrapping. However, hardest challenge is to make a balance between simplifying the complexity of decryption circuit as well as bootstrappability condition and the security of the underlying hard problems, which may require large parameters that leads to encryption schemes of high bit-complexity.

2.3 Few Related Works

Concept of homomorphic encryption scheme is expected to be a promising solution in the domain of cloud computing (Micciancio 2010; Boneh et al. 2013). Few previous works like Boneh et al. (2005) and Ishai and Paskin (2007) discuss about working on encrypted data, however Gentry's work in Gentry (2009a, b) propose the first plausible construction of FHE, which allows arbitrary computations on encrypted data. After Gentry's proposition, there are few other advancements in FHE literature mentioned in Stehle and Steinfeld (2010), Smart and Vercauteren (2010), Gentry-Halevi (2011a, b). Few notable works on fully homomorphic encryption over integers are investigated in contributions like van Dijk et al. (2009), Coron et al. (2011, 2012, 2014), Cheon et al. (2013). Some other contributions works like Brakerski et al. (2012) and Gentry et al. (2012a) focus on improving the performance of FHE scheme. Use of FHE in multi-party computation is further explored in López-Alt et al. (2012). Comparatively simpler homomorphic encryption scheme based on learning with errors has been reported in Gentry et al. (2013). Homomorphic evaluation of AES has been explored in Gentry et al. (2012b) and Doröz et al. (2014). Lattice based FHE scheme has been investigated in Brakerski and Vaikuntanathan (2014b). Some recent developments for faster FHE has been proposed in Alperin et al. (2014), Doröz et al. (2014) and Ducas and Micciancio (2015).

Classifications of different FHE contributions in literature based on different hard problems have been mentioned in Fig. 2.2. Based on these different schemes, few libraries are proposed in literature to extend real world problems in encrypted domain. Different libraries and their underlying mathematical background are detailed in Appendix D. Other than libraries, designing encrypted homomorphic compilers helps to put a step forward to achieve real world encrypted computing. Table 2.1 highlights few encrypted compilers with their brief background.

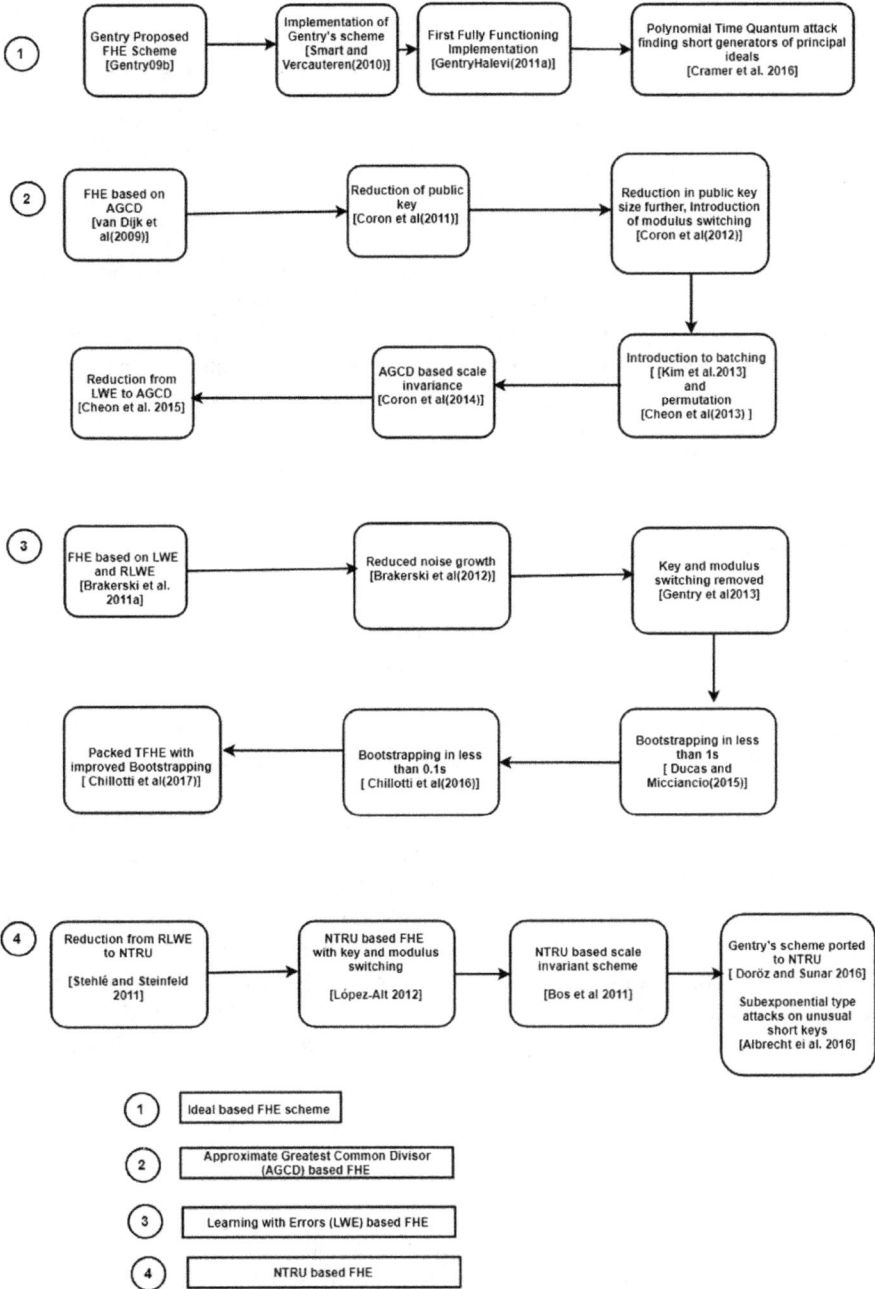

Fig. 2.2 FHE research Evolution Martins et al. (2018)

Table 2.1 Encrypted compilers in literature

Compiler	Scheme
Cingulata	• Open-source compiler toolchain
	• C++ programming interface for writing applications
	• Implementation of Fan–Vercauteren cryptosystem Cingulata (2018)
ALCHEMY	• Modular domain-specific language (DSL) for homomorphic computations on ciphertexts
	• No FHE domain knowledge is required for programmers Crockett et al. (2018) ALCHEMY (2018)
Standard API	• Storage Model API to represent data in cryptographic context
	• HE assembly language for circuit description information
	• Programming Model for business logic and circuit compilation
[for RLWE based HE]	Brenner et al. (2018)

2.4   FHE in Practical Algorithms

Most of the works reported in literature emphasize on performance improvement of FHE. Initial efforts have been made to investigate the answer of one pertinent question is *how far it is practical to execute secret program using fully homomorphic encryption* as mentioned in the work Brenner et al. (2011, 2012a). As a homomorphic operation, possibility of encrypted search operation has been investigated in Brenner et al. (2012b), Perl et al. (2012, 2014). Another work in Zhou and Wornell (2014) explores efficient homomorphic encryption on integer vectors and its applications. However, the paper does not deal with FHE supporting any arbitrary computations. *CryptDB* Popa et al. (2011, 2014) is an example of similar platforms that allows users to search and sort encrypted data. However, CryptDB is already known to be insecure solution shown in the work Akin and Sunar (2014). The work in Choudhury et al. (2013) explains the advantage of using FHE to make secure multi-party computation practical. With all these developments till date, we explore in detailed manner the feasibility and challenges of executing arbitrary algorithms when data is encrypted with classic FHE scheme.

2.5   Conclusion

Mentioned existing works in present literature are more about theoretical improvement of FHE performance. It does not discuss the impact of realizing arbitrary practical algorithms on FHE data. In subsequent chapters, we highlight that although FHE provides the capability of bitwise encrypted addition and multiplication, how it leaves several challenges for executing algorithms which run over the same. In the next chapter, we consider the example of sorting and analyze the challenges of realizing different sorting algorithms on encrypted data.

References

Akin IH, Sunar B (2014) On the difficulty of securing web applications using cryptdb. In: 2014 IEEE fourth international conference on big data and cloud computing, BDCloud 2014, Sydney, Australia, 3–5 Dec 2014, pp 745–752

ALCHEMY, https://github.com/cpeikert/ALCHEMY. Accessed 11 Oct 2018

Alperin-Sheriff J, Peikert C (2014) Faster bootstrapping with polynomial error. In: CRYPTO. Springer, pp 297–314

Boneh D, Gentry C, Halevi S, Wang F, Wu DJ (2013) Private database queries using somewhat homomorphic encryption. Springer, Berlin, pp 102–118

Boneh D, Goh E, Nissim K (2005) Evaluating 2-dnf formulas on ciphertexts. In: Proceedings of the theory of cryptography, second theory of cryptography conference, TCC 2005, Cambridge, MA, USA, 10–12 Feb 2005, pp 325–341

Brakerski Z, Gentry C, Vaikuntanathan V (2012) (leveled) fully homomorphic encryption without bootstrapping. In: Innovations in theoretical computer science, pp 309–325

Brakerski Z, Vaikuntanathan V (2014) Lattice-based FHE as secure as PKE. In: Innovations in theoretical computer science, ITCS'14, Princeton, NJ, USA, 12–14 Jan 2014, pp 1–12

Brenner M, Dai W, Halevi S, Han K, Jalali A, Kim M, Laine K, Malozemoff A, Paillier P, Polyakov Y, Rohloff K, Savas E, Sunar B (2017) A standard api for rlwe-based homomorphic encryption. HomomorphicEncryption.org, Redmond WA, Technical report

Brenner M, Perl H, Smith M (2012a) How practical is homomorphically encrypted program execution? an implementation and performance evaluation. In: Proceedings of the 11th IEEE international conference on trust, security and privacy in computing and communications, TrustCom 2012, Liverpool, United Kingdom, 25–27 June 2012, pp 375–382

Brenner M, Perl H, Smith M (2012b) Practical applications of homomorphic encryption. In: SECRYPT 2012 - Proceedings of the international conference on security and cryptography, Rome, Italy, 24–27 July 2012, pp 5–14

Brenner M, Wiebelitz J, Voigt G, Smith M (2011) Secret program execution in the cloud applying homomorphic encryption. In: Proceedings of 5th IEEE international conference on digital ecosystems and technologies (IEEE DEST 2011), pp 114–119

Bugiel S, Nürnberger S, Sadeghi AR, Schneider T (2011) Twin clouds: secure cloud computing with low latency. In: Proceedings of the 12th IFIP international conference on communications and multimedia security, CMS'11, pp 32–44

Cheon JH, Coron J, Kim J, Lee MS, Lepoint T, Tibouchi M, Yun A (2013) Batch fully homomorphic encryption over the integers. In: Advances in cryptology - EUROCRYPT 2013, 32nd annual international conference on the theory and applications of cryptographic techniques, pp 315–335

Choudhury A, Loftus J, Orsini E, Patra A, Smart NP (2013) Between a rock and a hard place: interpolating between MPC and FHE. In: Advances in cryptology - ASIACRYPT 2013 - 19th international conference on the theory and application of cryptology and information security, Bengaluru, India, 1–5 Dec 2013, Proceedings, Part II, pp 221–240

Cingulata, https://github.com/CEA-LIST/Cingulata/wiki. Accessed 11 Oct 2018

Coron JS, Mandal A, Naccache D, Tibouchi M (2011) Fully homomorphic encryption over the integers with shorter public keys. In: Proceedings of the 31st annual conference on advances in cryptology, pp 487–504

Coron J, Lepoint T, Tibouchi M (2014) Scale-invariant fully homomorphic encryption over the integers. In: Public-key cryptography - PKC 2014 - 17th international conference on practice and theory in public-key cryptography, pp 311–328

Coron J, Naccache D, Tibouchi M (2012) Public key compression and modulus switching for fully homomorphic encryption over the integers. In: Advances in cryptology - EUROCRYPT 2012 - 31st annual international conference on the theory and applications of cryptographic techniques, Cambridge, UK, 15–19 April 2012. Proceedings, pp 446–464

Doröz Y, Hu Y, Sunar B (2014) Homomorphic AES evaluation using NTRU. IACR Cryptology ePrint Archive

Ducas L, Micciancio D (2015) FHEW: bootstrapping homomorphic encryption in less than a second. In: Advances in cryptology - EUROCRYPT 2015 - 34th annual international conference on the theory and applications of cryptographic techniques, Sofia, Bulgaria, 26–30 April 2015, Proceedings, Part I, pp 617–640

Eric Crockett, Chris Peikert, Chad Sharp (2018) ALCHEMY: a language and compiler for homomorphic encryption made easy. ACM Conf Comput Commun Secur 2018:1020–1037

Fraleigh JB (2002) First course in abstract algebra. Addison-Wesley, Boston

Gamal TE (1985) A public key cryptosystem and a signature scheme based on discrete logarithms. IEEE Trans Inf Theory 31(4):469–472

Gennaro R, Gentry C, Parno B (2010) Non-interactive verifiable computing: Outsourcing computation to untrusted workers. In: Proceedings of the 30th annual conference on advances in cryptology, CRYPTO'10. Springer, Berlin, pp 465–482

Gentry C (2009a) A fully homomorphic encryption scheme. PhD thesis, Stanford University

Gentry C (2009b) Fully homomorphic encryption using ideal lattices. In: Mitzenmacher M (ed) STOC. ACM, pp 169–178

Gentry C, Halevi S (2011a) Fully homomorphic encryption without squashing using depth-3 arithmetic circuits. In: IEEE 52nd annual symposium on foundations of computer science, FOCS 2011, Palm Springs, CA, USA, 22–25 Oct 2011, pp 107–109

Gentry C, Halevi S (2011b) Implementing gentry's fully-homomorphic encryption scheme. In: Advances in cryptology - EUROCRYPT 2011 - 30th annual international conference on the theory and applications of cryptographic techniques, Tallinn, Estonia, 15–19 May 2011. Proceedings, pp 129–148

Gentry C, Halevi S, Smart NP (2012a) Better bootstrapping in fully homomorphic encryption. In: Public key cryptography - PKC 2012 - 15th international conference on practice and theory in public key cryptography, pp 1–16

Gentry C, Halevi S, Smart NP 2012b Homomorphic evaluation of the AES circuit. In: Advances in cryptology - CRYPTO 2012 - 32nd annual cryptology conference, pp 850–867

Gentry C, Sahai A, Waters B (2013) Homomorphic encryption from learning with errors: Conceptually-simpler, asymptotically-faster, attribute-based. In: Advances in cryptology - CRYPTO 2013 - 33rd annual cryptology conference, pp 75–92

Goldreich O, Ostrovsky R (1996) Software protection and simulation on oblivious RAMs. J ACM 43(3):431–473. https://doi.org/10.1145/233551.233553

Goldwasser S, Kalai Y, Popa RA, Vaikuntanathan V, Zeldovich N (2013a) Reusable garbled circuits and succinct functional encryption. In: Proceedings of the Forty-fifth annual ACM symposium on theory of computing, STOC '13. ACM, New York, NY, USA, pp 555–564

Goldwasser S, Micali S (1982) Probabilistic encryption & amp; how to play mental poker keeping secret all partial information. In: Proceedings of the Fourteenth annual ACM symposium on theory of computing, STOC '82, pp 365–377

Ishai Y, Paskin A (2007) Evaluating branching programs on encrypted data. In: TCC 2007

Kolesnikov V, reza Sadeghi A, Schneider T (2009) How to combine homomorphic encryption and garbled circuits improved circuits and computing the minimum distance efficiently

López-Alt A, Tromer E, Vaikuntanathan V (2012) On-the-fly multiparty computation on the cloud via multikey fully homomorphic encryption. In: Proceedings of the 44th symposium on theory of computing conference, STOC, pp 1219–1234

Martins P, Sousa L, Mariano A (2018) A survey on fully homomorphic encryption: an engineering perspective. ACM Comput Surv 50(6): 83:1–83:33

Melchor CA, Gaborit P, Herranz J (2008) Additively homomorphic encryption with d-operand multiplications. IACR Cryptology ePrint Archive

Micciancio D (2010) A first glimpse of cryptography's holy grail. Commun ACM 53(3):96

Paillier P (1999) Public-key cryptosystems based on composite degree residuosity classes. In: Proceedings of the 17th international conference on theory and application of cryptographic techniques, EUROCRYPT'99, pp 223–238

Perl H, Mohammed Y, Brenner M, Smith M (2012) Fast confidential search for bio-medical data using bloom filters and homomorphic cryptography. In: eScience. IEEE Computer Society, pp 1–8

Perl H, Mohammed Y, Brenner M, Smith M (2014) Privacy/performance trade-off in private search on bio-medical data. Future Generation Computer Systems, pp 441–452

Popa RA (2014) Building practical systems that compute on encrypted data. PhD thesis, Massachusetts Institute of Technology

Popa RA, Redfield Catherine MS, Zeldovich N, Balakrishnan H (2011) Cryptdb: Protecting confidentiality with encrypted query processing. In: Proceedings of the 23rd ACM symposium on operating systems principles, SOSP '11, pp 85–100

Rass S, Slamanig D (2013) Cryptography for security and privacy in cloud computing. Artech House Inc, Norwood

reza Sadeghi A, Schneider T, Win M (2010) Token-based cloud computing secure outsourcing of data and arbitrary computations with lower latency. Workshop on trust in the cloud

Rivest RL, Adleman L, Dertouzos ML (1978a) Foundations of Secure Computation. On data banks and privacy homomorphisms. Academia Press, Cambridge, pp 169–179

Sahai A (2008) Computing on encrypted data. In: ICISS. Springer, pp 148–153

Sander T, Young AL, Yung M (1999) Non-interactive cryptocomputing for nc1. In: FOCS. IEEE Computer Society, pp 554–567

Smart NP, Vercauteren F (2010) Fully homomorphic encryption with relatively small key and ciphertext sizes. In: Proceedings of the 13th international conference on practice and theory in public key cryptography, PKC'10, pp 420–443

Stallings W (2005) Cryptography and network security, 4th edn. Prentice-Hall Inc, Upper Saddle River

Stehle D, Steinfeld R (2010) Faster fully homomorphic encryption. Cryptology, ASIACRYPT 2010:377–394

UniCrypt, https://github.com/bfhevg/unicrypt/blob/master/README.md. Accessed 11 Oct 2018

van Dijk M, Gentry C, Halevi S, Vaikuntanathan V (2009) Fully homomorphic encryption over the integers. IACR Cryptology ePrint Archive

Yao AC (1982) Protocols for secure computations. In: Proceedings of the 23rd annual symposium on foundations of computer science, SFCS '82, pp 160–164

Zhang Z (2014) Revisiting fully homomorphic encryption schemes and their cryptographic primitives. PhD thesis, University of Wollongong

Zhou H, Wornell GW (2014) Efficient homomorphic encryption on integer vectors and its applications. In: 2014 Information theory and applications workshop, ITA 2014, San Diego, CA, USA, 9–14 Feb 2014, pp 1–9

Chapter 3
Sorting on Encrypted Data

Sorting is an age old problem of rearrangement. Since arrangement of items has pro-
found influence on speed and simplicity of algorithms, sorting has attracted great deal
of importance in computer science literature. Recently with the advent of cloud computing we revisit problem of sorting on encrypted data. Sorting network[1] consists of
comparators and swapping operations. The difference between classical comparison-
based sorting algorithms and sorting networks on encrypted inputs is that all operations must be data independent in the flow of the algorithm steps in sorting networks.
Hence, in spite of the fact that data dependent algorithms may be faster, they may
not suitable for encrypted data.

Encrypted sorting is the basic building block of *ORDER BY* operation. Sorting,
that in turn requires encrypted comparison is the main pillar in this context. There
are different works in literature to perform encrypted data comparisons. Yao's circuit
evaluation Yao (1982) was used by Vaidya and Clifton (2003) for the comparisons in
their privacy preserving k-means clustering algorithm. Secure comparison protocol
proposed by Fischlin (2001) uses GM-homomorphic encryption scheme from the
work Goldwasser and Micali (1982) and AND homomorphic encryption conversion
technique from XOR homomorphic encryption in GM scheme by Sander et al. (1999).

In the work Liu et al. (2014), authors propose the idea of order preserving indexing
to compare encrypted ciphertext. The data owner (client) uses a trap-door to compute
the index which is randomized and provided to the service provider (cloud server).
The randomization prevents the leakage of the trap-door. However, as the authors
mention that if FHE schemes proposed in Naehrig et al. (2011) are used, the order

---

[1]In computer science, comparator networks are abstract devices built up of a fixed number of
"wires", carrying values, and comparator modules that connect pairs of wires, swapping the values
on the wires if they are not in a desired order. Such networks are typically designed to perform
sorting on fixed numbers of values, in which case they are called sorting networks (Sorting Network
2018).

© Springer Nature Singapore Pte Ltd. 2019
A. Chatterjee and K. M. M. Aung, *Fully Homomorphic Encryption
in Real World Applications*, Computer Architecture and Design Methodologies,
https://doi.org/10.1007/978-981-13-6393-1_3

preserving indices will not remain randomized well. Furthermore, the fact that the service provider can determine the relative ordering of plaintexts from ciphertexts can have potential security implications (Stinson 2002). In present literature, there are few more works on encrypted search, however contributions on encrypted sorting are still very limited. In Çetin et al. (2015), few sorting techniques on somewhat homomorphic encrypted data have been discussed. work mentioned in Baldimtsi and Ohrimenko (2015) concentrates on specialized framework of structured data. Sorting and searching schemes in this mentioned paper use additive homomorphic encryption as an underlying scheme with some predefined assumptions and shows overall performance improvement. In Damgård et al. (2007), a protocol has been proposed for secure comparison in multi party comparison framework considering players are honest but curious and the underlying homomorphic encryption is again partial homomorphic. Proposed scheme in Zhou and Wornell (2014) do not work on FHE data and CryptDB (Popa et al. 2011) is also vulnerable to certain attacks (Akin et al. 2015). In Emmadi et al. (2015) authors have mentioned theoretically odd-even sort or bitonic FHE sort as an efficient sorting technique, but have not justified how such algorithms have been implemented without suitable parallel processor. Another important sorting technique has been discussed as oblivious sorting of secret shared data, that is effective in case of multi-party computation (Bogdanov et al. 2014). However, for cloud servers where arbitrary data is stored encrypted with FHE schemes (to support other arbitrary operations in encrypted domain.), dedicated techniques are required for performing specific operations. In this context, this chapter explores different sorting techniques to find which one is suitable to apply on FHE datasets. Fully homomorphic circuits for performing comparison based swaps has been proposed and used to realize conventional sorting algorithms (Table 3.1).

Table 3.1 Types of sorting techniques (for n items)

Sorting	Time-complexity [Average Case]	Memory requirement
Parition based sorts		
Quicksort	$nlogn$	in-place with $O(logn)$ stack space
Mergesort	$nlogn$	$O(n)$
Heapsort	$nlogn$	$O(1)$
Comparison based sorts		
Insertion sort	n^2	$O(1)$
Selection sort	n^2	$O(1)$
Bubble Sort	n^2	$O(1)$

3.1 FHE Comparison Based Sort

Comparison based sorting algorithms are based on conditional swap operations. When data is encrypted, this operation need to be translated to Fully Homomorphic Swap (FHS) operation. Any FHE library supports bit-wise encrypted addition and multiplication. Hence, we shall discuss how to use the following Fully Homomorphic primitive circuits to realize the next-level complex FHS circuit:

- *FHE_ADD*: Add ciphertexts.
- *FHE_MUL*: Multiply ciphertexts
- *FHE_Fulladd*: Add with carry in and carry out
- *FHE_Halfadd*: Add with carry out

The FHS circuit depends on two main operations: subtraction operation and decision making based on the subtraction result. Fully Homomorphic subtraction, which is implemented by performing homomorphic addition of one ciphertext with 2's complement of another ciphertext. For two plaintext numbers a and b, subtraction can be computed as:

$$a - b = a + 2\text{'s complement of b} \tag{3.1}$$

Now a homomorphic subtraction of a' and b' which are the encryptions of a and b respectively is computed using the homomorphic addition as follows:

$$a' - b' = a' + Encrypt(2\text{'s complement of b}) \tag{3.2}$$

The 2's complement of b in the encrypted domain is obtained as follows:

$$Encrypt((2's\ complement\ of\ b), pk) = b' \oplus Encrypt(11\ldots1, pk) \oplus Encrypt(1, pk)$$

Figure 3.1 represents addition module in fully homomorphic domain and it is used to design the subtraction circuit as shown in Fig. 3.2. The MSB of the subtraction

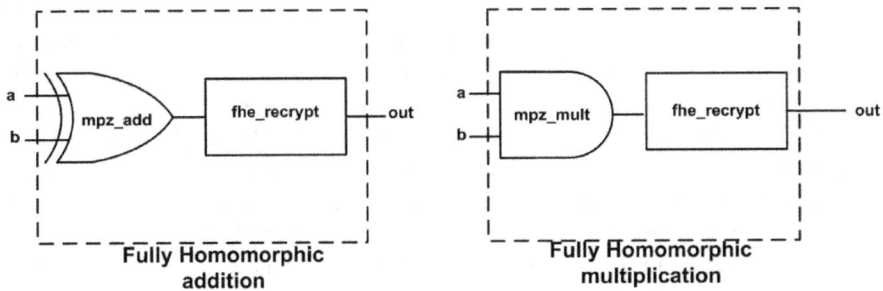

Fig. 3.1 Fully homomorphic addition and multiplication (Chatterjee and SenGupta 2013)

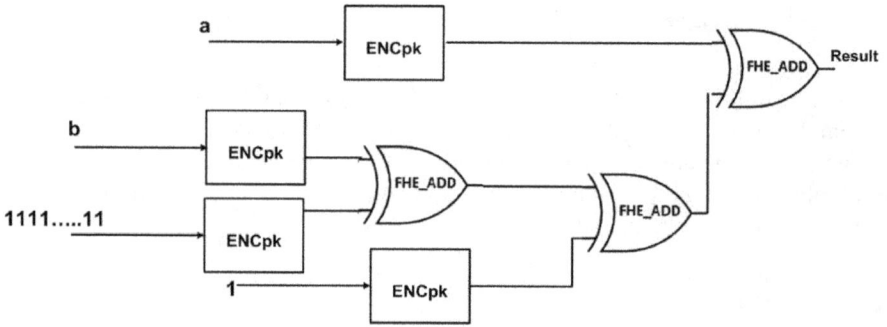

Fig. 3.2 Fully homomorphic subtraction (Chatterjee and SenGupta 2013)

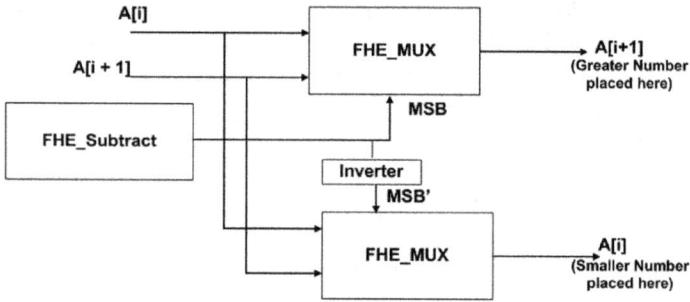

Fig. 3.3 Fully homomorphic swap (Chatterjee and SenGupta 2013)

output is further fed to the decision making module as a selection line. The following equations represent how the swap operation takes place between two elements $A[i]$ and $A[i + 1]$ depending on MSB (represented here as bt):

$$temp = bt * A[i] + (1 - bt) * A[i + 1]$$
$$A[i + 1] = (1 - bt) * A[i] + bt * A[i + 1]$$
$$A[i] = temp$$

Figure 3.3 shows the overall swap operation in fully homomorphic domain. As the figure depicts, other than the fully homomorphic subtraction, FHS operation also depends on decision making of multiplexer(MUX) in homomorphic domain.

Figure 3.4 represents the fully homomorphic MUX (*FHE_MUX*) designed with *FHE_ADD* and *FHE_MULT* modules. In the figure, a' and b' are the encryptions of the two inputs a and b and MSB represents the most significant bit of the subtraction output. In every stage, $Encrypt(1, pk)$ need to be added to invert a bit using *FHE_ADD* circuit. Finally, the output result of the MUX is computed as follows:

$$\overline{(a'.MSB).\overline{(b'.\overline{MSB})}} = a'.MSB + b'\overline{MSB} \qquad [applying\ De\ Morgan's\ law]$$

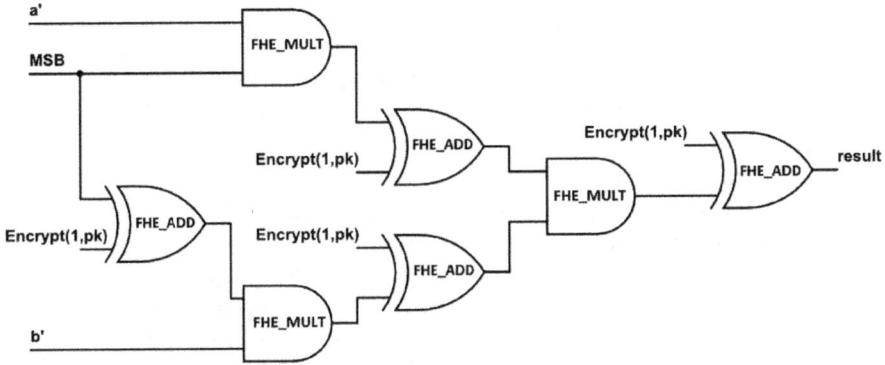

Fig. 3.4 Fully homomorphic 2:1 Mux (Chatterjee and SenGupta 2013)

Algorithm 1: Encrypted Bubble sort

Input: Unsorted FHE encrypted array *enc_arr*
Output: Sorted FHE encrypted array *enc_arr*
int i, j, k, m;
$mpz\_t * temp1$; $char * ptext$;
for $(i = 0; i < lenarr - 1; i + +)$ **do**
    for $(j = 0; j < lenarr - i - 1; j + +)$ **do**
        $copy\_mpz\_arr(temp1, enc\_arr[j])$;
        $copy\_mpz\_arr(temp2, enc\_arr[j + 1])$;
        $fhe\_Swap(enc\_arr[j + 1], enc\_arr[j], temp1, temp2, pk)$;

$\overline{var}$ represents the bit inversion of *var*, where *var* is any variable. The overall module is designed with the in hand functions available to Library libScarab (2011). With the help of this encrypted swap operation, sorting operation in homomorphic domain has been developed.

3.1.1 Homomorphic Form of Sorting

In this section, we start our discussion with the most common sorting algorithm, Bubble sort. In Algorithm 3.1 we present an overview of fully homomorphic bubble sort.

To implement the algorithm, unencrypted variables should be mapped to encrypted datatype specific to the underlying library (used *mpz_t* data-type is mentioned in the Library libScarab 2011). This variable type is library dependent and will vary according to the choice of libraries. The function *copy_mpz_arr* is to copy *enc_arr[j]* to the temp array. *fhe_Swap* is the function for above mentioned FHS operation and responsible for the main swap operation in the comparison based sorting. The sorting was performed on input sizes of length 5–40 for repeated runs of around 20 times and sorting of 40 data takes around 21565 s. The mentioned algorithm has been evaluated for correctness on a Linux Ubuntu 64-bit machine 1.6 GHZ clock and 8 GB RAM. Experiments in Chatterjee and SenGupta (2013) show that sorting of around 50 integers hit maximum RAM limit of 8 GB and consequently sorting aborts due to huge ciphertext size expansion.

In our next discussion, it will be explored whether it is really possible to achieve performance gain by partition based sort (over encrypted comparison sort) maintaining the security of the cryptosystem. Hence, in the next section the feasibility of implementing partition based sort on FHE data will be analyzed in the light of security and will be explained the proposed scheme mentioned in Chatterjee and SenGupta (2017) to actually implement partition sort on encrypted data.

3.2 Sorting and Security

In this section, we shall investigate the security implications of performing partition based sort on FHE data. Basically, we examine whether the FHE scheme remains secured against *Chosen-Plaintext Attacks (CPA)* (will be discussed in Sect. 3.2.1) if it is possible to perform partition based sort on encrypted data. Considering Quick sort as an example of partition based sort, encrypted quick sort will be discussed in the subsequent section. Quick sort algorithm requires three steps:

1. Selection of pivot element.
2. Comparison of each element with the pivot and decision making to partition.
3. Perform sorting recursively on each of the partition the array.

Now, we shall explain quick sort in the light of CPA indistinguishability experiment following the set-up explained in Katz and Lindell (2007) and show how the possibility of performing encrypted partitioning may lead to the cryptosystem being vulnerable to CPA attacks.

3.2.1 The CPA Indistinguishability Experiment

Considering any public-key encryption scheme $\Pi = (Gen, Enc, Dec)$, an adversary Adv and security parameter n, the following steps explain CPA indistinguishability experiment $PubK_{Adv,\Pi}^{cpa}(n)$:

1. A key k is generated by $Gen(n)$. The key comprises of a public key and private key.
2. Adv is given input n and oracle access to $Enc_k(.)$, and outputs a pair of messages m_0, m_1 of the same length.

Fig. 3.5 CPA due to
partition based FHE sorting
(Chatterjee and SenGupta
2017)

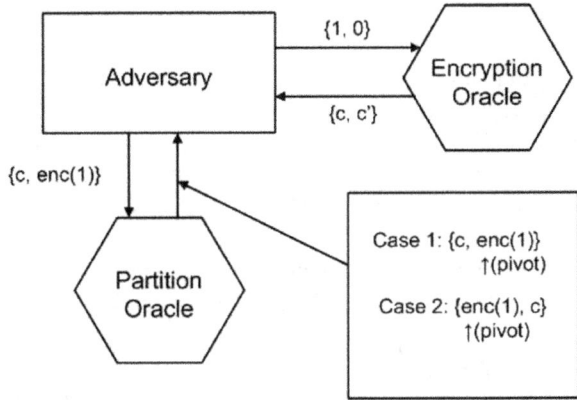

3. A random bit $b \leftarrow 0, 1$ is chosen, and then a ciphertext $c \leftarrow Enc_k(m_b)$ is computed and given to Adv. Here, c is the *challenge ciphertext*.
4. The adversary Adv can use the oracle access to $Enc_k(.)$ to output a bit b'.
5. The output of the experiment is defined to be a success, denoted by 1 if $b' = b$, and a failure indicated by 0 otherwise. Thus, if $PubK_{Adv,\Pi}^{cpa}(n) = 1$, Adv is successful.

Thus, as mentioned in Stinson (2002) a public-key encryption scheme $\Pi = (Gen, Enc, Dec)$ has indistinguishable encryptions under a chosen-plaintext attack (or is CPA secure) if for all probabilistic polynomial-time adversaries A there exists a negligible function $negl$ such that: $Pr[PubK_{A,\Pi}^{cpa}(n) = 1] \leq \frac{1}{2} + negl(n)$.

More formally the situation will be explained with Fig. 3.5 and the following steps:

1. Let an adversary Adv be capable of performing partition based sort on an encrypted database D and returns the final sorted array.
2. Figure 3.5 shows that that in such scenario adversary Adv has access to a *partition oracle* and an *encryption oracle*. *Encryption oracle* returns the encryption of any message, once the message is given as an input. The *Partition oracle* takes as input an array of encrypted data and a pivot element, and returns two partitions of the input array. One partition contains all the elements lesser than the pivot element and other partition holds all elements equal to or greater than the pivot element.
3. Now, the adversary Adv receives a challenge c which is the encryption of either of the messages m_0 or m_1. The adversary wins if he is capable of guessing correctly whether c is the encryption of m_0 or m_1. Adversary has to guess the bit b with a probability non-negligibly higher than $1/2$, where $Enc(pk, m_0) = c$ (where pk is the public key). For convenience, $m_0 = 0$ and $m_1 = 1$. Thus, $c = Enc(b)$ where $b \in \{0, 1\}$.

Now, any successful adversary Adv, which uses a partition oracle can obtain partition based sorted array on encrypted data (without any single decryption). However, that indicates the decision of partitioning (whether one encrypted data is lesser or greater compared to encrypted pivot) is not hidden to the Adv. In this scenario, adversary can send $\{c, Enc(pk, 1)\}$ to the partition oracle and make it as pivot. Now, the partition oracle can return two results:

Case 1: $\{c, \underbrace{Enc(1)}_{\text{pivot}}\}$ indicates c is lesser than the pivot (note that ordering is with

respect to plaintext) and clearly it is the encryption of 0.

Case 2: $\{\underbrace{Enc(1)}_{\text{pivot}}, c\}$ indicates c is equal to or greater than the pivot and clearly it

is the encryption of 1.

Based on the response of the partition oracle, the adversary is thus capable of easily guessing the bit b. The plaintext 0 or 1 which resulted in the challenge ciphertext c. Hence, it is evident that the cryptosystem is prone to chosen plaintext attack if it is possible to perform partition based sorting on encrypted data in the traditional way of comparison with pivot (as done on unencrypted data). It is interesting to note in the next section that the comparison based sort is not vulnerable to the CPA attack.

3.2.2   Why Comparison Based Sorting is Secured?

Comparison based sort on FHE data is performed using fully homomorphic conditional swap (FHS) operation. The FHS circuit depends on two main operations: subtraction and homomorphic multiplexing. It is interesting to note that the capability to perform the above partition based sort can lead to a CPA adversary, while such a reduction is not possible for a comparison based sort. Close observation of the homomorphic swap operation explains why this sorting does not lead to the CPA adversary. The crux of this lies in the fact that it is never disclosed to the adversary which of the two inputs of the *FHE_Swap* (FHS) block is greater. Since, all the elements that are fed as the input to the swap circuit are modified by the FHE operations by the FHE primitive circuits, changes also take place in the output ciphertext, ensuring that they correspond to the same plaintexts. The security of the FHE scheme is retained since an adversary is incapable of collecting the information whether the swap operation is really taking place or not. Note that the adversary cannot tally the outputs with the inputs, as they are changed and the equality is only in the plaintexts which is hidden to the attacker. This makes comparison based sort secured.

The basic difference between comparison based and partition based sorting lies in their working techniques. Comparison sort works fine in the encrypted domain because it is not required to know if one given input is homomorphically greater (or smaller) than another input, but still one can place the greater elements and the smaller elements in their correct positions. However, this information is necessary in the partitioning phase of quick sort. In this phase of quick sort, one compares the elements of the array with the pivot and increases or decreases the running indices based on the results of the comparisons. However, in homomorphic domain this is not feasible as the FHS step never reveals whether the swap took place, and also changes the values of the inputs. Hence, the array of encrypted numbers cannot be partitioned in the classical sense. On the contrary, the ability to partition in the classical sense implies that the underlying FHE data can be subjected to a CPA attack.

In the next section, we take a fresh look at partitioning and make an attempt to realize quick sort on FHE data. To be more precise, the limitation in applying the partitioning to FHE can be alleviated if one also encrypts the index of the array. This layer of encryption helps one to hide the position of the pivot and thus although the partitioning happens the exact pivot index is hidden to the adversary. In the next section, we detail the idea of index encryption and explain how to use it to realize quick sort.

3.3 Partition Based Sorting with Index Encryption

In this section, we shall discuss the steps to perform partition based sort on an array of FHE data. The essential function for performing quick sort is as follows:

```
void quickSort(array, first, limit)
{
    if (first < last)
    {
        p = partition(A, first, last);
        /* Partitioning index */
        quickSort(A, first, p - 1);
        quickSort(A, p + 1, last);
    }
}
```

The above code follows standard recursion using which quick sort on unencrypted data can be implemented. The essential functions are partition and two recursive calls to the quick sort routine itself. Recursive codes are realized on the system stack, which performs two operations push and pop to maintain the intermediate operations of the program. Now consider the situation when the array is encrypted. As discussed, the position of the pivot is also computed in an encrypted fashion, and thus the limits of the arrays (which are the indices) are also stored in an encrypted fashion. However, the system stack is unable to realize the above recursion. This is because the stack stores the start (left) and the end (right) indices in an encrypted format. But the decision to pop or push (to decrement or increment the stack pointer) depends on the encrypted pivot position and the left or right index, both of which are also encrypted. Thus the address of the stack also needs to be encrypted, and hence one needs to develop a user-defined encrypted stack to realize a FHE quick sort. In the following, we provide an overview starting with an organization of the encrypted array with an encrypted index.

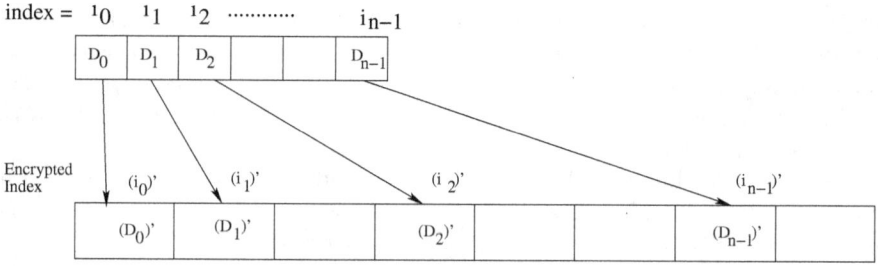

Fig. 3.6 Encrypted array with encrypted indices (Chatterjee and SenGupta 2017)

3.3.1 Encrypted Array with Encrypted Index

Figure 3.6 shows how an array A is mapped to its encrypted form. The base address of A is encrypted (by FHE encryption) and that is considered as the base address of the encrypted array. The next encrypted locations are obtained by performing FHE additions to the previous locations. For example, enc(1) is added (by FHE addition) to the encrypted ith location address (i') and encrypted address of $(i + 1)$th location is obtained. However, due to the randomness property of FHE encryption, the consecutive unencrypted addresses do not remain consecutive after encryption. Finally, data of ith location (say D_i) is placed to i' location as $(D_i)'$ in the encrypted array.

This encrypted array is chosen as the data structure for implementing partition based quick sort on encrypted data. Now, deciding the implementation approach of the algorithm is the next challenge. Recursive and iterative are the two different practiced implementation approaches of performing quick sort. Here, we first discuss what are the problems of implementing recursive methods on encrypted data and then explain proposed scheme handling such implementation challenges.

3.3.2 Problems of Recursion on Encrypted Data

In general, recursive implementations are very popular for partition based sort on unencrypted data from design point of view. However, it is required to specify the initialization or termination condition of recursion. While working with encrypted data and encrypted indices, the initialization or termination conditions of recursion are the results of encrypted FHE operations (by homomorphic modules). However, present underlying processors are unencrypted and unable to process such encrypted recursion conditions. Hence, direct implementation of recursion methods are not possible in such existing processors.

In case of recursive implementation of partition sort, intermediate partition parameters are stored in recursion stack. For handling encrypted sort, an encrypted auxiliary stack based design is explained in Chatterjee and SenGupta (2017) and encrypted

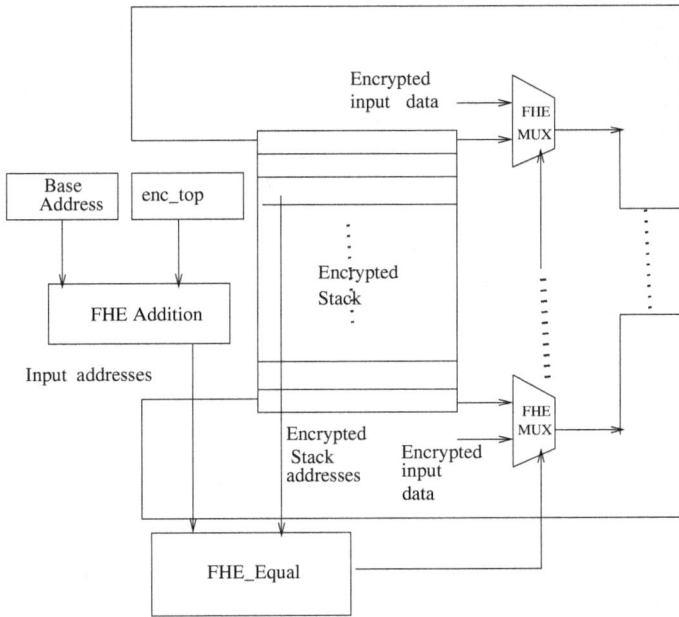

Fig. 3.7 Encrypted push operation (Chatterjee and SenGupta 2018)

partition indices values are stored in this encrypted stack. Now, we outline the design of an encrypted stack which is capable of handling encrypted push and pop operation and extend the idea on how to use it implementing partition based sort. Encrypted stack can be defined with following operations:

(a) Encrypted Push Operation

Push is responsible for both initialization and data insertion to stack. For initialization of stack, an address is encrypted and that is considered as the starting base address of stack. Stack size (mentioned as encrypted enc_top) is $Enc(0)$ at this point.

During data insertion (Push operation), the encrypted base address of stack is increased by enc(1) each time and encrypted data is stored in the next address. enc_top is increased accordingly to hold the index of the top of the stack (which indicates the size of the stack too). All these increment operations are again homomorphic and take place using FHE addition operations. Here enc_top is added with encrypted base address value and data is pushed when address match is found. Figure 3.7 explains the push operation to an encrypted stack. Every stack location is basically defined as structure having encrypted data and encrypted address as the two elements.

(b) Encrypted Pop Operation

Pop operation in such encrypted stack is even more interesting since all encrypted data are now residing in encrypted addresses. Figure 3.8 explains the pop operation from an encrypted stack. During this operation, the enc_top decreases and (when

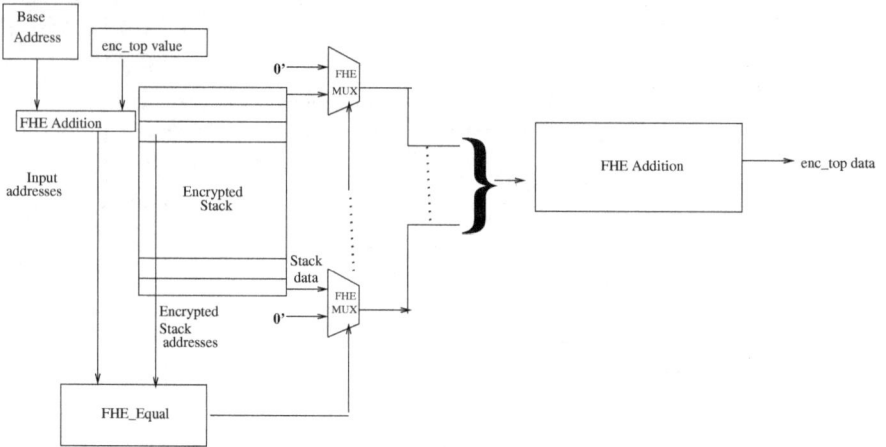

Fig. 3.8 Encrypted pop operation (Chatterjee and SenGupta 2018)

added to the base address) gives the modified address from where data should be popped. However, the challenge is this modified address value is not exactly bit-wise same with the existing stack locations. This creates the real ambiguity in case of fetching (or popping) data from an encrypted stack. Here we use the encrypted multiplexer to check which encrypted stack location is homomorphically equal to the modified address. If a match is found, data is popped from that particular location. However, encrypted multiplexer gives an encrypted result (address matched or not) and hence the location (from which data is actually fetched) remains hidden from any adversary. Here enc_top is added with encrypted base address value and data is popped when address match is found.

(c) Error Handling

During handling of any stack, two error conditions are possible. Firstly, attempting to push data when the stack is full and the secondly trying to pop data when the stack is empty. Both these error conditions can be handled by comparing the existing stack size with maximum stack size (to check whether the stack is full) or with $Enc(0)$ (to check whether the stack is empty).

This encrypted stack is used to store the start and end indices of the partition and to continue quick sort in sub-arrays.

3.3.3 Quick Sort Using Encrypted Stack

In this section, we discuss the encrypted quick sort where data is encrypted and stored in an array with encrypted index locations. The FHE quick sort function has the following arguments:

```
fhe_qsort(encarray, encfirst, enclimit, pk)
```

The arguments are respectively the encrypted array, the encrypted starting address of the array, and the encrypted end address of the array and the public key *pk*. As discussed the implementation is done using an encrypted stack, which stores the intermediate partition parameters. Partition function is described as follows.

3.3.3.1   Encrypted Partitioning

The partition function takes as input the encrypted array, along with the encrypted left and right indices. The initial pivot can be chosen as the first, middle or any random element of the input array. However, the last index (encrypted) has been chosen here. The final position of the pivot is determined by encrypted comparisons. A crucial step of the partitioning algorithm is to compare homomorphically the encrypted pivot *pv* data with the encrypted data pointed by an encrypted running index, j'. Based on the comparison result, another encrypted running index i', is incremented.

This comparison is done using the comparison circuit *FHE_isgreater* (will be detailed in Chap. 4), which operates on two encrypted values to check if one value is greater than the other. The two encrypted values are subtracted and MSB of the subtraction result is fed as the input of the comparison circuit. If the MSB is Enc(1), the first input is lesser than the second, else otherwise. The following code snippet shows how encrypted partitioning takes place using the encrypted operations:

```
fhe_isgreater(cond_loop, enclimit, jIndex, pk);
fhe_isgreater(cond2, data_pivot,  data_indexj, pk);
fhe_mul(cond, cond_loop, cond2, pk);
fhe_mux(mod_i, iIndex,  iIndex_incr, cond, pk);
```

The comparison result between `enclimit` and the encrypted index `jIndex` homomorphically sets the value of a variable `cond_loop` depending on the result of the comparison. The encrypted bit `cond_loop` decides whether i' will be incremented or hold the previous value. Likewise, the encrypted value stored in the encrypted address `jIndex`, denoted as `data_indexj`, is compared with the data of the encrypted pivot, `data_pivot` and another condition, `cond2` is set homomorphically. Finally, both the conditions are homomorphically multiplied to generate the signal `cond`. The update of the index i' is done through a homomorphic multiplexer, denoted as `fhe_mux` which takes the encrypted value of i, `iIndex` and the $(i + 1)'$, `iIndex_incr` as inputs. The selection among the two inputs is done by the encrypted value `cond`.

Thus selection of the proper index continues until the correct position of the pivot is determined in the encrypted form. However, since all the decisions are performed in the encrypted format, one cannot decide on the termination condition without decrypting. Hence, the pivot based partitioning, although performed homomorphically from the left index of the array (or sub-array in the recursive calls), to the right index, is always executed on the complete array length of maximum number of data to be sorted. This is a major bottleneck in terms of performance, and overall time complexity.

3.3.4 Encrypted Quick Sort Implementation

The overall sorting will be performed using a user-defined stack, where both the content and the address are encrypted. The stack stores the encrypted end limits of the array portions currently being sorted (now onwards `encfirst` and `enclimit` will be mentioned as l and h). Encrypted Push operation is responsible for both initialization and data insertion of the encrypted array limits to the stack. For initialization of the stack, an address is encrypted and that is considered as the starting base address of stack. Stack size (mentioned by encrypted stack_top) is $Enc(0)$ at this point.

The sorting function at each iteration pops out the end limits of the encrypted array. The partition function (discussed in the previous section) provides the encrypted pivot position p. Subsequently, the encrypted array indices $p - 1$ and l are homomorphically compared to check whether $p - 1 > l$ and if true, the stack is pushed with the at the locations l and $p - 1$. Note the decision to push the stack depends on an encrypted comparison, and thus implying why the stack addresses also needs to be encrypted. Likewise, encrypted index $p - 1$ will be pushed and it is checked whether $p + 1 < h$, and in a similar manner push the indices $p + 1$ and h. Both the push and pop stack operations are encrypted as mentioned in Sect. 3.3.2.

In the quick sort, push or pop operation to or from the stack continues till $enc\_top \geq 0$. However, implementation of this step (in encrypted quick sort) requires a comparison between encrypted $stack\_top$ and $Enc(0)$, which in turn generates an encrypted result. To solve this issue, the entire loop is run for the size n of the array, where n data is being sorted (again termination cannot be done depending on the stack being empty). The result is still functionally correct, as there will be redundant operations which does not make any change to the array in the homomorphic sense. Likewise for partitioning, each time the comparison check is done over the entire array length. However, for a given call to the partition function, sorting happens from the running index l to h (the current left and right indices of the array). For ranges outside this limit, no change or sorting is done in the homomorphic sense. This of-course has an adverse effect on the time complexity, and in fact provides bad timing than the comparison based sorts, because of increased stack operations. However, as discussed, this redundancy is mandatory since without decryption one cannot reveal the index position of the stack or determine the termination condition. Finally, the proposed sorting algorithm on encrypted data meets the security requirement as explained in Sect. 3.2 since all the performed intermediate operations are encrypted and an encrypted sorted array is produced as the final result. In the next section, we formally analyze the timing requirements of proposed sorting schemes (comparison as well as partition based schemes) to decide which encryption scheme is most suitable while working on encrypted data.

3.4 Timing Requirement for Sorting Schemes on Encrypted Data

In this section, comparison and partition based sorting techniques on encrypted data are compared in terms of their time complexities for n data stored in the cloud and encrypted with FHE scheme. To perform encrypted comparison sort on this database we need to have n encrypted comparisons in each of the n iterations as explained in Sect. 3.1. Hence, as a worst case analysis, it requires n^2 comparisons. Let the time for each encrypted comparison be T_c and the total time for comparison sort T_{comp} can be computed as $T_{comp} = n^2 * T_c$.

For performing quick sort on encrypted data as explained in Sect. 3.3, partitioning and stack handling are the two main operations. Let the partitioning time be T_p and the stack handling time be T_s and hence for each iteration the sorting time is $T_p + T_s$. Hence, the total time requirement for quick sort is $T_{quicksort} = n * (T_p + T_s)$.

Now, for each partition operation let the main array is divided into two sub-arrays. Now, each of the elements of each partition need to be compared with pivot, but the comparison loop should iterate for n times (maximum length of main input array) for each partition sub-array (since the actual partition index is encrypted and actual partition length is not known). Hence, for each iteration partition time becomes $T_p = 2 * n * T_c$ and the overall time becomes:

$$\begin{aligned}
T_{partition} &= n * (T_p + T_s) \\
&= n * (2 * n * T_c + T_s) \\
&= 2 * n^2 * T_c + n * T_s
\end{aligned} \tag{3.3}$$

Hence, the time complexity of quick sort on encrypted data is no better than comparison sort. Further, the stack operation adds extra overhead to this timing requirement. Since, the required time for encrypted push and pop operations proportionally increases with the increase of the stack size, this stack timing overhead T_s is also very high as shown in actual timing requirements of the sorting schemes presented in Table 3.2.

Table 3.2 Comparison of different sorting techniques (Chatterjee and SenGupta 2017)

Sorting	No. of elements	Average time required (s)
Bubble sort	5	235
	10	1527
	40	21565
Quick sort	5	776
	10	4102
	40	46757

3.4.1 Performance Analysis of Different Operations

Table 3.2 shows required time for performing different sorting with increasing number of data. To investigate the actual reason of performance hindrance, while working with FHE data timing for different sub-operations are analyzed for 32 bit size integers (shown in Table 3.3).

Using these values the required time for sorting have been theoretically estimated which mainly depends on time for each comparison, which in turn again depends on FHE addition, multiplication and recrypt operation. For example, total timing requirement for n^2 comparisons, where $n = 40$ is around $((40^2) * 26.5/2)$ s or 21,200 s. The value is very close to practical time as observed in Table 3.2. Figure 3.9 shows the comparison between theoretical and practical requirements.

The timing requirement in Table 3.3 confirms that FHE recrypt, which is the main reason of large timing overhead for operations in homomorphic domain. Hence, the ways to improve the performance of FHE sorting time will be investigated by reducing recrypt. However, reduction of recrypt below a certain level can introduce error and abolishes the homomorphic property of FHE data. In the next section, it

Table 3.3 Estimate of different sub-operations for sorting (Chatterjee and SenGupta 2013)

	Operation	Time required (s)
1.	FHE add	0.4
2.	FHE mul	0.5
3.	FHE swap(greater/smaller)	14.2
4.	FHE swap(greater+smaller)	26.5
6.	FHE recrypt	0.4

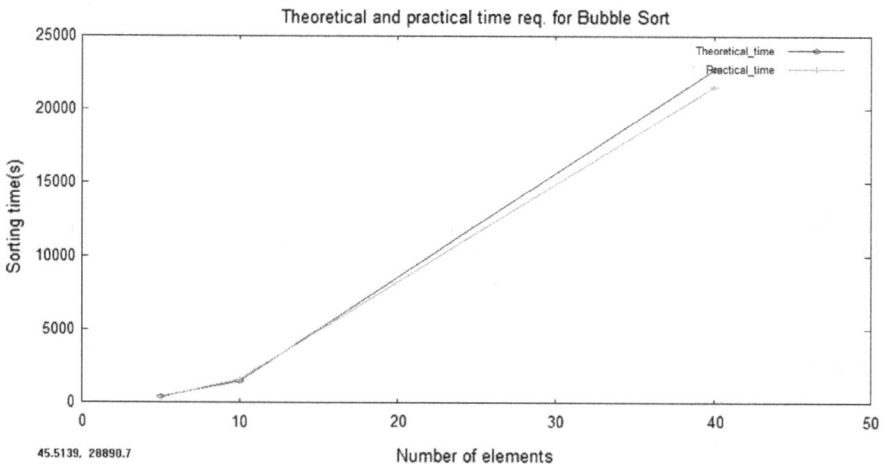

Fig. 3.9 Timing requirement for fully homomorphic bubble sort (Chatterjee and SenGupta 2013)

is explored how the presence of error can be used as an advantage to performance improvement and a new kind of FHE data specific sorting is introduced termed as Lazy sort.

Lazy Sorting

In this section we discuss the concept of lazy-sorting which is based on lazy FHE swap operation which was proposed in Chatterjee and SenGupta (2017). Çetin et al. (2015) and Emmadi et al. (2015) have mentioned LazySort does not prove to be better over insertion sort contradicting the claim of Chatterjee and SenGupta (2017). However, it is not clear from contributions of Çetin et al. (2015) and Emmadi et al. (2015) how they have actually implemented insertion sort. The limitation of implementing encrypted insertion sort on unencrypted processor will be explained in detail in the following section to explain Lazy Sort further. In this operation, huge number of unencrypted data has been analyzed to check how much erroneous swaps can be tolerated to result in an almost sorted array. This almost sorted array is further converted to final sorted array but with an advantage of less number of recrypts. An average of several experiments with more than 1000 data to measure the allowable error shows that on an average with around 30% error, around 60% data are placed in proper position.

Minimization of recrypt after a certain threshold may introduce some error in the comparison decision(swap) of fully homomorphic encrypted data. Hence, this will in turn introduce some error in the sorting decision. The term *error* indicates an element is placed in wrong position in the final sorted array. Now, if any erroneous comparison sort is performed with minimized recrypt, it will take comparatively less time due to the use of comparison circuit with reduced number of recrypts and results an almost sorted array. Finally, insertion sort can be applied theoretically, which works in linear time for an almost sorted array.

3.4.2 Further Reduction of Recrypt to Introduce Error

It is evident that it is not possible to remove all the recrypts since it will introduce 100% error and result in positioning large number of elements in the wrong place in the sorted array. For this reason, careful choice of removable recrypts are necessary. Identified recrypts are present in FHE subtract module, which is one of the main submodule of FHE swap operation. In this function, the recrypt operations required to correct the values of carry bit of the addition result in *FHE_Fulladd* module are removed. This in turn reduces the time requirement in every iteration of addition and finally reduces the time for swap operation. However, use of this modified swap operation results correct output with 70% accuracy. This results in an almost sorted array. Subsequently, we apply insertion sort, which has linear time complexity for almost sorted array. Thus, Lazy Sort is composed of two steps:

1. Initially, comparison based Bubble sort is performed with erroneous swap operations, where error is introduced by minimizing recrypt operation. Since, recrypt

is the costly operation in terms of timing requirement, reduction of recrypt may enhance the performance. However, the sorting operation results an almost sorted array.

2. In the final stage, insertion sort with encrypted swap is applied, since it works in linear time on an almost sorted array.

However, handling insertion sort is not also straightforward while working with encrypted data. Here, we take a code-snippet for insertion sort and discuss the challenge of implementation.

```
#include<stdio.h>
int main(){

1.    int i,j,n,temp,a[20];
2.    for(i=0;i<s;i++)
          scanf("%d",&a[i]);

3.    for(i=1;i<n;i++){
4.        temp=a[i];
5.        j=i-1;
6.        while((temp<a[j])&&(j>=0)){
7.        a[j+1]=a[j];
8.            j=j-1;
9.        }
10.        a[j+1]=temp;
11.    }
12.    return 0;
}
```

In the above code-snippet, Lines 3 – 9 shows the main comparison block for insertion sorting. To sort n FHE data, the intermediate comparisons should be encrypted. It is assumed that the information about how much data is getting sorted, hence unencrypted value of s is expected to be known and implementation of loop iteration of line 3 is straightforward.

However, difficulty lies in handling the while loop with all encrypted operations, since, temp and a[j] are encrypted and generates encrypted results after (temp < a[j]) operation. However, all existing underlying processors are unencrypted, hence they are incapable of handling encrypted loop conditions. Thus, in this scenario it is impossible to identify the termination point of encrypted while loop. This explains why it is infeasible to implement insertion sort on encrypted data.

In the next section, we shall discuss a special window based technique to actually implement the encrypted insertion sort following Chatterjee and SenGupta (2015). Since, insertion sort is supposed to be applied on an almost sorted array, the fact is used that in an almost sorted array, the correct position of the element lies within a window size w. This indicates, for an almost sorted array a, an element whose correct position is supposed to be $a[i]$ can lie within position $a[i + w]$ with probability 1. It is intuitive that the value of w will depend on the error in the comparisons, which in turn depends on the number of recrypt operations. In next section, an estimate will be provided on the window length and justify that the probability of an element to erroneously lie outside the window falls exponentially with the value of w.

Choosing Suitable Window Size

In this section, a relationship will be established between possible window size (w) and probability of residing an element in a certain position after erroneous propagation. an array a is considered to compute the probability of an element being wrongly placed in position $i + w$ with error probability p, where $(0 < i < w)$. Table 3.4, tabulates all the notations used in this proof.

Starting with window size 2, if $a[1]$ is erroneously placed to $a[2]$ with error probability p, then

$$P_{12} = [Pr(a[1] > a[2]).Pr(error)] = \frac{1}{2}.p \tag{3.4}$$

For a window size 3, probability of an element to propagate from $a[1]$ to $a[3]$ is $P_{12} + P_{23}$.

Similarly, let P_{13} be the probability of element $a[1]$ erroneously propagating to $a[3]$. In this case, P_{13} can be computed as:

$$\begin{aligned}
P_{13} &= P_{12} + P_{23} \\
&= (P_{12E}.P_{23NE}) + (P_{12E}.P_{23E}) + (P_{12NE}.P_{23E}) \\
&= \frac{1}{2}.p.\frac{1}{2}.(1 - p) + \frac{1}{2}.p.\frac{1}{2}.p + \frac{1}{2}.(1 - p).\frac{1}{2} \\
&= \frac{1}{2}.p - \frac{1}{4}.p^2 \tag{3.5}
\end{aligned}$$

For obtaining a general term, erroneous propagation is considered of an element from n^th location to $(n + 1)^th$ location. Probability of erroneous propagation will be

Table 3.4 Notations in the proof of concept

Notations	Explanations
P_{ij}	Probability of an element a[i] is propagating to a[j]
P_{ijE}	Probability of an element a[i] is erroneously propagating to a[j]
P_{ijNE}	Probability of an element a[i] is propagating to a[j] without error
p	Probability of erroneous comparison

considered up to n as P_{1n}. From n to $n + 1$, the propagation can be erroneous or error free. Hence, following two scenarios are important:

1. Erroneous propagation from 1 to n, then final propagation from n to $n + 1$ can be erroneous or error-free.
2. Error free propagation from 1 to n and then error is introduced within the propagation from n to $n + 1$.

Thus, the overall probability is computed as:

$$P_{n+1} = P_n \cdot \frac{1}{2} + \left[\frac{1}{2}(1 - p) \right]^{(n-1)} \cdot \frac{1}{2} \cdot p \tag{3.6}$$

Thus if an window size w is considered such that $(w < n)$, the probability of erroneous propagation of an element P_{1w} is:

$$P_{1w} = \frac{1}{2^w} + \left[\frac{1}{2} \cdot (1 - p) \right]^{(w-1)} \cdot \frac{1}{2} \cdot p \tag{3.7}$$

For an optimum size of w, $\frac{1}{2} \cdot (1 - p)^{(w-1)} \cdot \frac{1}{2} \cdot p$ term becomes negligible. Hence, value of P_{1w} becomes $\frac{1}{2^w}$.

Hence, it can be concluded when an erroneous comparison occurs, the probability of an element (which is supposed to be placed in $a[i]$) to be placed in $a[w]$ reduces as w increases. Hence, it can be assumed for an optimum window size, the element is expected to be present within the window in spite of having some erroneous comparison. this assumption can further be used for implementing insertion sort on FHE data.

In the next section, we discuss what are the practical challenges of implementing this two stage sorting on FHE data and show how assumption of placement of an element within a window can make the implementation of encrypted insertion sort actually feasible.

3.4.3 Encrypted Insertion Sort

We follow the code-snippet of insertion sort explained in Sect. 3.4.2 and explain how the challenge explained in the above section can be handled to implement the sorting with FHE operations. The `for` loop in line 3 is implemented as an unencrypted loop, since it is known how many elements s are getting sorted. In line 4, encrypted `a[i]` is assigned to `temp` and in line 5, `j-1` is performed by FHE subtraction. However, as explained in Sect. 3.4.2, the main challenge is to implement `while((temp < a[j])&&(j >= 0))` of line 6.

One possible solution of implementing this `while` loop is to realize it as an unencrypted `for` loop. Since, `temp` and `a[j]` both are encrypted, it should be

Table 3.5 Window size versus error probability (Chatterjee and SenGupta 2015)

Window size	Error probability
5	0.5
10	0.1
12	0.02
15	0.06

hidden which `a[j]` is satisfying the condition (`temp < a[j]`). Hence, the entire `for` loop can iterate over the maximum possible length (which can be the maximum possible number of elements `s`). However, this way of implementation does not give any advantage of using insertion sort if the data are almost sorted, since each time the inner loop iterates overall length of data (`s`) and the complexity of the sorting remains $O(s^2)$, for s number of elements. Hence, a window based technique is applied to implement insertion sort on almost sorted array.

Table 3.5 shows a comparative result of error probabilities with different window sizes. In the next section, we discuss how this window based insertion sort is incorporated in the Lazy sort.

3.4.3.1 Window Technique Based Insertion Sort

In this section, we focus on implementation of a modified insertion sort, which is advantageous while applying on almost sorted array. We follow the code-snippet of traditional insertion sort in which Line 6 shows basic insertion step. To insert any new element `a[i]` (which is stored as `temp`), the element is compared with all other elements of previous locations ((`j= i-1`) and lesser) and this comparison continues until `a[j]` is smaller. However, difficulty with FHE data is underlying unencrypted processor cannot understand where this loop is getting terminated. Hence, it is impossible to identify actual insertion point of `a[i]`, while working with encrypted data. Hence, one possible solution is to iterate the `while` for whole length of data. Here, one assumption about insertion sort is applied while working on almost sorted array. Since, the array is almost sorted it is assumed that that insertion position of the next element is not absolutely random and expected to be inserted within a certain window length w from the position (`i-1`). Hence, it is sufficient to execute a loop of length w and it is expected that proper insertion position should belong within this window length (Fig. 3.10).

3.4.3.2 Window Based Insertion Sort in Lazy Sort

Finally, this modified insertion sort is applied in the second stage of Lazy sort. In the initial stage, bubble Sort is performed with reduced recrypt operation which introduces some error in the final sorted array and generates an almost sorted array. In the next stage, this window based technique of insertion sort is applied on almost sorted

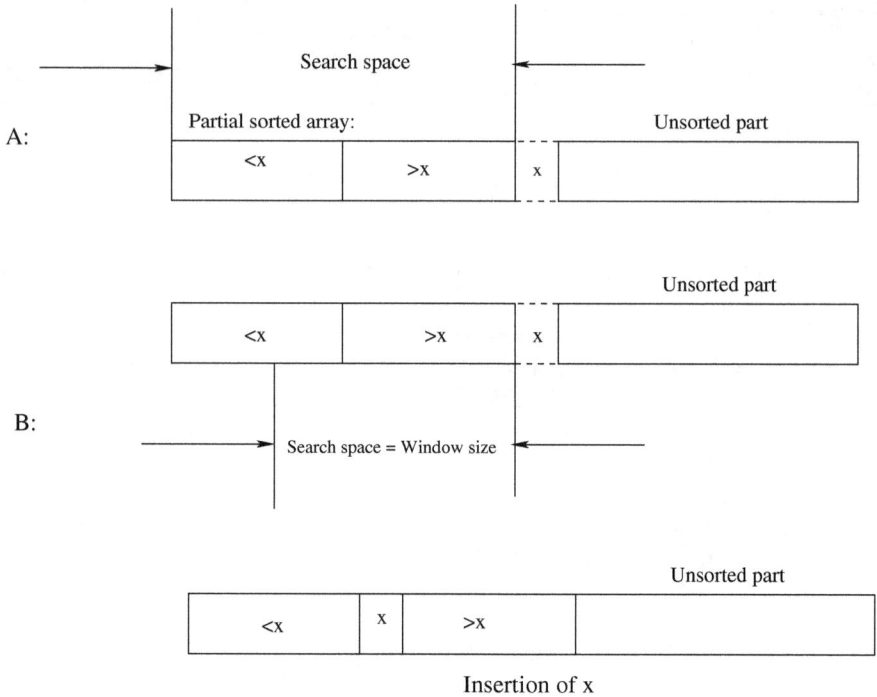

Fig. 3.10 Case 1: Traditional insertion Case 2: Window based insertion (Chatterjee and SenGupta 2017)

array. The code snippet below shows how the `while` loop of unencrypted insertion sort (lines 3–9) are realized with FHE operations and window based technique:

```
for(i=1;i<n;i++)
{
    temp' = a'[i];
    j = i-1;
    for(cnt = 0; cnt< w; cnt++){
        fhe_isgreater(grt, a'[j], temp');
        fhe_mux (arrvaltemp, temp', a'[j], grt);
        a'[j+1] = arrvaltemp;
        j=j-1;
    }
}
```

This code snippet shows how to execute encrypted insertion sort on FHE array a'. Encrypted variable `temp'` stores value of `a'[i]` and `j=i-1`. When *w* is chosen as window size, each time `a'[j]` is compared with `temp'` with `fhe_isgreater` module. If `a'[j]` is greater than `temp'` value of `grt` is set `enc(1)`. Depending

Table 3.6 Timing analysis for lazy sort (Chatterjee and SenGupta 2017)

Number of data	Fully homomorphic sorting time (s)	Lazy sort time (s)		
		$W = 2$	$W = 6$	$W = 10$
10	1527	923	1623	2150
20	5300	3092	4292	5492
30	11980	6507	8307	9065
40	21750	11168	13544	15972

on the value of grt, next value of a'[j+1] is set as temp' or a'[j] decided by encrypted multiplexer fhe_mux module. Thus, a'[i] is placed in the proper position.

However, this window based technique depends on the specific assumption that the proper insertion position of the elements will be within the window. Hence, there is a possible error probability in applying this window based technique in loop handling of insertion sort. As we have discussed how this error can be minimized with suitable choice of optimum window size (an optimum window size $w = 10$ is chosen here to get the finally sorted array). Time complexity of the initial erroneous Bubble sort is $O(n^2)$. However, intermediate operations of bubble sort are performed with reduced recrypt operation, hence the timing requirement is less. Further, the time complexity of next stage window based insertion sort is $O(nw)$, which is almost $O(n)$ considering $w << n$ for large number of FHE data in practical scenario. Thus, this two stage sorting technique enhances the performance of encrypted sorting technique as shown in Table 3.6.

3.5 Conclusion

In this chapter, we discuss techniques for sorting which can be effective on encrypted data. Fully homomorphic addition and multiplication operations are slow due to the presence of cipher refreshing recrypt technique. This in turn deteriorates the overall performance. With this motivation, it was experimented to find the effect of eliminating recrypts, the error arising thereof and the effect of that on sorting. Finally, based on these observations, two stage lazy-technique for sorting is mentioned which improves the time for sorting significantly. This faster two stage encrypted sorting named Lazy sort is implemented by an initial erroneous Bubble sort with reduced recrypt followed by insertion sort which works in linear time on almost sorted array. Further. challenges of implementing insertion sort is highlighted on FHE and proposed window based method that helps to implement the insertion sort on encrypted data in a practical way to make this two stage sorting actually feasible. The classical work in Knuth (1998) has concluded that "the bubble sort seems to have nothing to recommend it, except a catchy name and the fact that it leads to some interesting theoretical problems." However, illustrations in this chapter show that bubble sort has much to offer while performing secure computation on encrypted data. As a future work, various ways of improving sorting performance can further be explored by applying parallelization and other techniques in software domain and perform a formal timing analysis for modified sorting schemes. In the next chapter, we extend the focus to explore how arbitrary algorithms can be realized using FHE operations.

References

Akin IH, Sunar B (2015) On the difficulty of securing web applications using CryptDB. IACR cryptology ePrint archive, vol 2015, p 82

Baldimtsi F, Ohrimenko O (2015) Sorting and searching behind the curtain. In: Financial cryptography and data security - 19th international conference, FC 2015, San Juan, Puerto Rico, 26–30 January 2015, Revised selected papers, pp 127–146

Bogdanov D, Laur S, Talviste R (2014) A practical analysis of oblivious sorting algorithms for secure multi-party computation. In: Bernsmed K, Fischer-Hübner S (eds) Secure IT systems. NordSec 2014. Lecture notes in computer science, vol 8788. Springer

Çetin GS, Doröz Y, Sunar B, Savaş E (2015) Low depth circuits for efficient homomorphic sorting. Cryptology ePrint Archive, Report 2015/274

Chatterjee A, Kaushal M, SenGupta I (2013) Accelerating sorting of fully homomorphic encrypted data. INDOCRYPT 262–273

Chatterjee A, SenGupta I (2015) Windowing technique for lazy sorting of encrypted data. In: 2015 IEEE conference on communications and network security (CNS), pp 633–637

Chatterjee A, SenGupta I (2017) Sorting of fully homomorphic encrypted cloud data: can partitioning be effective? IEEE Trans Serv Comput

Chatterjee A, SenGupta I (2018) Translating algorithms to handle fully homomorphic encrypted data on the cloud. IEEE Trans Cloud Comput 287–300

Damgård I, Geisler M, Krøigaard M (2007) Efficient and secure comparison for on-line auctions. Springer, Berlin, pp 416–430

Emmadi N, Gauravaram P, Narumanchi H, Syed H (2015) Updates on sorting of fully homomorphic encrypted data. IACR cryptology ePrint archive, vol 2015, p 995

Fischlin M (2001) A cost-effective pay-per-multiplication comparison method for millionaires

Goldwasser S, Micali S (1982) Probabilistic encryption & how to play mental poker keeping secret all partial information. In: Proceedings of the Fourteenth annual ACM symposium on theory of computing, STOC '82, pp 365–377. ACM, New York . http://doi.acm.org/10.1145/800070.802212

Katz J, Lindell Y (2007) Introduction to modern cryptography (Chapman & Hall/CRC Cryptography and network security series). Chapman & Hall/CRC

Knuth DE (1998) The art of computer programming, vol 3, 2nd edn. Sorting and searching. Addison Wesley Longman Publishing Co., Inc, Redwood City, CA, USA

Library libScarab (2011). https://github.com/hcrypt-project/libscarab

Liu D, Bertino E, Yi X (2014) Privacy of outsourced k-means clustering. In: A 9th ACM symposium on information, computer and communications security, ASIA CCS '14, Kyoto, Japan - June 03 -06 2014, pp 123–134

Naehrig M, Lauter K, Vaikuntanathan V, Can homomorphic encryption be practical? In: Proceedings of the 3rd ACM workshop on cloud computing security workshop, ser. CCSW '11. New York, NY, USA: ACM, pp 113–124

Popa RA, Redfield Catherine MS, Zeldovich N, Balakrishnan H (2011) CryptDB: protecting confidentiality with encrypted query processing. In: Proceedings of the 23rd ACM symposium on operating systems principles (SOSP), Cascais, Portugal

Sander T, Young A, Yung M (1999) Non-interactive cryptocomputing for nc1. In: 40th annual symposium on foundations of computer science, pp 554–566

Stinson D (2002) Cryptography: theory and practice, second edition : section 5.9.1, 2nd edn. CRC/C&H

Vaidya J, Clifton C (2003) Privacy-preserving k-means clustering over vertically partitioned data. In: Proceedings of the Ninth ACM SIGKDD international conference on knowledge discovery and data mining, KDD '03, pp 206–215. ACM, New York. http://doi.acm.org/10.1145/956750.956776

Wikipedia, Sorting network (2018) [Online. accessed 22-July-2018]. https://en.wikipedia.org/wiki/Sorting_network

Yao AC (1982) Protocols for secure computations. In: Proceedings of the 23rd annual symposium on foundations of computer science, SFCS '82, pp 160–164. IEEE Computer Society, Washington, DC. http://dx.doi.org/10.1109/SFCS.1982.88

Zhou H, Wornell GW (2014) Efficient homomorphic encryption on integer vectors and its applications. In: 2014 Information theory and applications workshop, ITA 2014, San Diego, CA, USA, 9–14 February 2014, pp 1–9

Chapter 4
Translating Algorithms to Handle Fully Homomorphic Encrypted Data

An algorithm is defined as a self-contained step-by-step set of operations to be performed to solve a particular problem. The concept of simplicity and elegance related to any algorithm informally appears in Knuth's saying "... *we want good algorithms in some loosely defined aesthetic sense. One criterion is the length of time taken to perform the algorithm...Other criteria are adaptability of the algorithm to computers...*" (Knuth 1973).

Hence, adaptability to the underlying platform is a critical parameter while dealing with any algorithm. Further, it is said "*An algorithm must be seen to be believed.*". Classical algorithms are mostly non-circuit based. Thus, they are not described in terms of logical gate level operators, like AND-OR gates or multiplexers. This leads to potential problems in direct translation of traditional algorithms on general-purpose computers when executed on FHE data, since FHE operations are realized by gate level operators. For example, recursion is a typical non-circuit technique for efficient implementation of algorithms, but cannot be directly realized using FHE schemes. Problems arise for the stack manipulation on general-purpose computers and handling of initialization and termination conditions, which are also encrypted. Further, loops and conditional statements cannot be directly handled via FHE operations. Thus one needs to develop suitable synthesis techniques and methodologies to handle algorithms which operate on encrypted data.

In this chapter, we shall focus on the following:

- Identifying the basic components of an algorithm to realize them in the encrypted domain along with mapping the unencrypted variables to a data structure corresponding to encrypted variables or ciphertexts.
- Targeting algorithms in both non-recursive and recursive version and discussing their realizations while operating in the encrypted domain. A program being a sequence of instructions can be converted to the encrypted domain by replacing the instructions by their homomorphic equivalents (Chatterjee and SenGupta 2018). The instructions often set values of control variables, which decide the control flow path in the program. The challenge remains that in the homomorphic counterpart,

© Springer Nature Singapore Pte Ltd. 2019
A. Chatterjee and K. M. M. Aung, *Fully Homomorphic Encryption in Real World Applications*, Computer Architecture and Design Methodologies, https://doi.org/10.1007/978-981-13-6393-1_4

these variables are also encrypted. We discuss mechanism of translating these constructs using FHE based decision making to operate on the encrypted control variables. Interestingly, these components do not leak to an adversary the outcome of the decision.

- Likewise, it will be discussed how to deal with recursive programs and to show that one cannot implement such schemes on a general purpose computer in a straight-forward manner. In order to implement such algorithms we need to have an encrypted-stack, where both the address and the contents of the stack are encrypted. Hence, synthesizing various data structures, like linked-lists, stacks, and queues on FHE data will be given a special focus.

General Algorithm Handling with Their Encrypted Variants

Any algorithm is comprised of sequences of specific mathematical and logical manipulations. Conventional program which run on unencrypted data has a long history. They are executed on general purpose computers to perform operations as required to solve a problem. However, algorithms on encrypted data bring forward many new challenges, some of which are enumerated below:

4.1 Challenges of Executing Encrypted Programs

1. *Handing of arithmetic operations*: The underlying ALU operations supported by general purpose computers are traditional computations, like addition, multiplication, equality, comparisons, bitwise operators etc. On the other hand, encrypted programs demand the requirement of FHE operations, which are complex operations. Hence, translation of the code from an unencrypted to an encrypted program should be tackled efficiently.
2. *Handling of conditional branches and detecting termination conditions*: The conditional operations are encrypted and hence branching and termination need to be handled depending on encrypted conditions.
3. *Supporting recursive programs*: Handling recursion with encrypted operations is another major challenge since existing recursion stacks are unencrypted.

Our main objective is to study of implementing algorithms on encrypted data. For this reason, implementations of basic operations need to be defined. In the next section, we discuss about the basic operations required to implement standard algorithms and then those operations will be realized on encrypted data.

4.2 Encrypted Variants of Basic Operators

An algorithm consists of a set of variables and instructions to be executed on them, with possible iterative executions. Figure 4.1 shows the steps of translating an unen-

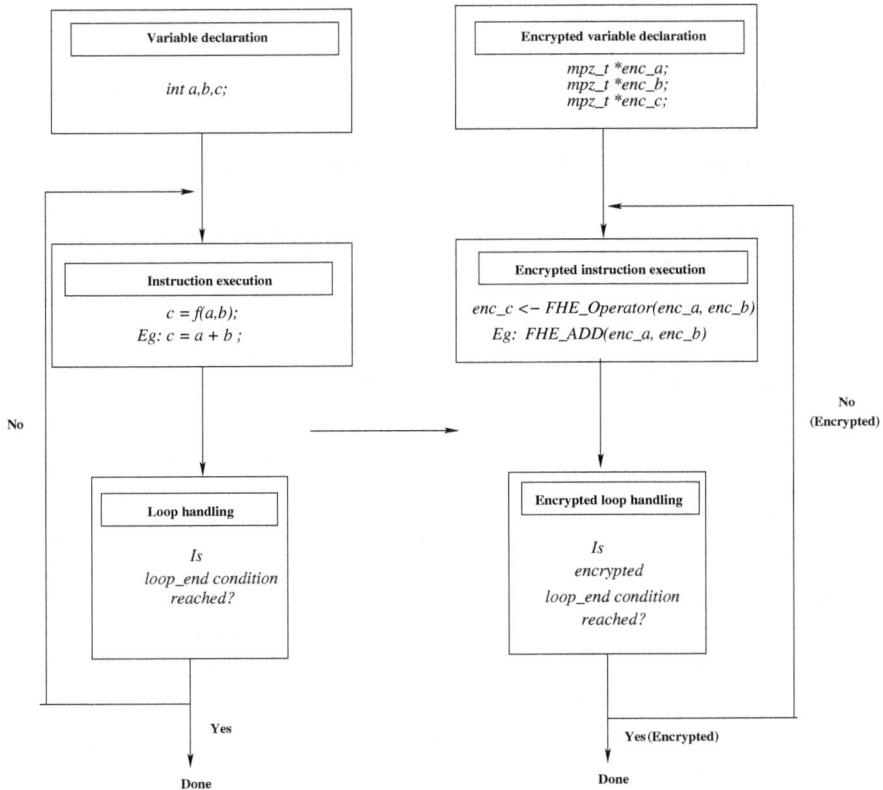

Fig. 4.1 Algorithmic steps (Chatterjee and SenGupta 2018)

crypted algorithm to its encrypted form. We elaborate each of the algorithm translation steps in our subsequent subsections.

The first step of translating any algorithm is realizing the variables in encrypted domain. For example, Library libScarab (2011) maps the integer input variables to *mpz_t* type encrypted variables. Instructions on the variables are basically a single operation or function to be executed as shown in Fig. 4.1. To handle those operations on encrypted data, we first identify the different class of unencrypted operators and then design their respective encrypted counterparts. According to the standard programming languages (like C), these manipulations or operations can be classified into different types:

- *Arithmetic Operators* perform addition, subtraction, multiplication and division of two operands.
- *Relational Operators* checks if the values of two operands are equal. These operators also decide the greater - lesser relations between two operands.
- *Logical Operators* include logical AND, OR and NOT operations among the operands.

- *Bitwise Operators* perform bitwise operations among the operands (like bitwise AND, OR etc).
- *Assignment Operators* assign values from right side expression to left side expression.

In the next subsection, we shall observe how these operations can be implemented on encrypted data. Initially, we explain how to handle the bitwise encrypted operations using circuit based FHE primitives, then subsequently other operations on FHE ciphertexts will be explained.

4.3 Encrypted Bitwise and Assignment Operators

FHE circuit based bitwise operations, like bitwise AND, OR operations on encrypted data are performed in the following way:

- Bitwise AND and OR can be usually performed by basic modules from FHE libraries.
- Bitwise inversion is performed by *FHE_INV* module, in which to invert an encrypted bit a', $Enc(1)$ is added to a' using *FHE_ADD* module. If the bit a' is $Enc(0)$, then addition with $Enc(1)$ inverts the bit to $Enc(1)$, else if the bit is $Enc(1)$, addition operation inverts it to $Enc(0)$.

All the logical operators are implemented by performing bitwise operation over the operand bit length. Encrypted assignment operation is comparatively straightforward, where one right encrypted operand value is copied to left encrypted operand value.

4.4 Encrypted Arithmetic Operators

In this section, implementation of encrypted arithmetic operations like encrypted addition, subtraction, multiplication and division are explained.

4.4.1 Encrypted Addition and Subtraction

In general, modules provided by FHE libraries perform addition between two encrypted bits. Final addition between two operands of size 32 bits can be extended by such FHE bitwise addition operations. Detailed design of FHE subtraction has been explained in previous chapter.

Algorithm 2: Shift and add algorithm

Input: $a = (a_{n-1}, a_{n-2} \ldots, a_0)_2$ and $b = (b_{n-1}, b_{n-2} \ldots, b_0)_2$
Output: product p
for $j \leftarrow 0$ *to* $n - 1$ **do**
    if $b_j = 1$ **then**
        $c_j = a$ shifted j places;
    else
        $c_j = 0$;

$p = 0$;
for $j \leftarrow 0$ *to* $n - 1$ **do**
    $p = p + c_j$;
return p;

4.4.2 Encrypted Multiplication

Typically, existing computers used *Shift and Add algorithm* to multiply unencrypted data as shown in Algorithm 4.1. In traditional shift and add multiplication, addition and shifting are both performed if the data-bit equals to 1. However, while working with encrypted data there is no scope to identify whether the bit is $Enc(0)$ or $Enc(1)$ due to the complete randomization of encryption scheme. If an adversary successfully differentiates between $Enc(0)$ and $Enc(1)$, then the cryptosystem is prone to chosen plaintext attack (Rass and Slamanig 2013). Hence, for two encrypted data A' and B', addition and shifting both the operations are performed for every bit (since, addition with $Enc(0)$ does not change the result in the corresponding computation with the plaintext, it is equivalent to one shifting). This incurs a significant overhead in terms of performance.

Further, direct application of such multiplication algorithm imparts certain limitations to actual implementation. For multiplication of two k-bit integers, the multiplication result grows to $2k$-bit in every stage. Next, we shall realize division on encrypted data.

4.4.3 Encrypted Division

Division algorithm computes quotient (Q) and remainder (R) of two given integers say (N) and (D). There are several categories of division algorithms on unencrypted data. Here, challenges of choosing division algorithm on encrypted data are investigated. Division algorithm are of two types:

- Slow division algorithms that produce one digit of the final quotient per iteration.
- Fast division methods that start with a close approximation to the final quotient and produce twice as many digits of the final quotient on each iteration.

Algorithm 3: Division algorithm by repeated subtraction

Input: Integers N and D
while $N \geq D$ **do**
$\quad \lfloor \; N \leftarrow N - D;$
return N;

Algorithm 4: Restoring division

Input: Integers N and D
Output: Quotient (Q) and remainder (R)
$P \leftarrow N;$
$D \leftarrow D << N;$
for $i \leftarrow n - 1$ *to* 0 **do**
$\quad P \leftarrow 2P - D;$
$\quad$ **if** $P >= 0$ **then**
$\quad\quad \lfloor \; q(i) \leftarrow 1;$
$\quad$ **else**
$\quad\quad q(i) \leftarrow 0;$
$\quad\quad P \leftarrow P + D;$

Table 4.1 Notations of the restoring division algorithm (Chatterjee and SenGupta 2018)

Notations	Explanations
N	Numerator
D	Denominator
n	number of bits
P	Partial remainder
q(i)	ith bit of quotient

Since ciphertext increases in size compared to the plaintext due to the property of encryption scheme, FHE fast division produces twice as many digits of the final quotient on each iteration. Hence, slow division method is suitable one for encrypted division. Typically, the simplest division algorithm is by repeated subtraction as shown in Algorithm 1.

However, the termination condition of this algorithm requires a step (**while** $N \geq D$). This step is impossible to handle by existing unencrypted processors since both N and D are encrypted and they produce an encrypted termination condition of the algorithm. Thus, restoring division algorithm is chosen on unencrypted data and translate it to its encrypted counterpart (Table 4.1).

Here are the following steps to translate the Algorithm 4 on encrypted data:

• Inputs N, D and all other intermediate variables are encrypted to mpz_t data-type.
• The addition and subtraction operations on encrypted data are implemented with FHE addition and FHE subtraction module as explained in Sect. 4.4.1.

- The greater than equal to operation ($P >= 0$) and the condition check if ($P >= 0$) are realized by FHE relational operator and FHE multiplexer. The details of these modules will be discussed later in Sect. 4.5.
- Finally, the (for ($i = n - 1 \ldots 0$)) loop iterates depending on the bit size of the inputs, which can be given as an unencrypted input to the program, since it does not leak any critical information. Hence, contrary to Algorithm 4.2, the advantage of selecting this algorithm is that the termination condition of the algorithm does not depend on any encrypted condition.

4.5 Encrypted Relational Operators

Relational operators are the decision making operators which decides if one operand is lesser (or greater) than the other operands. However, if any adversary gets the idea whether one ciphertext is less than the other in terms of plaintext, then the cryptosystem will be vulnerable to chosen plaintext attack. Hence, while working with encrypted data the homomorphic comparisons are performed using FHE operations and the generated results also remain encrypted.

4.5.1 Encrypted Comparison Operation

Figure 4.2 shows the *FHE_GrtEq* module to compute greater than equal to relation between two encrypted operands. The steps of the computation are stated below:

1. FHE subtraction is performed between two inputs (say a' and b').
2. MSB of the subtraction result is fed to an encrypted multiplexer (FHE_MUX) as the selection line. MSB equals to 1 indicates a' is less than equal to b', else

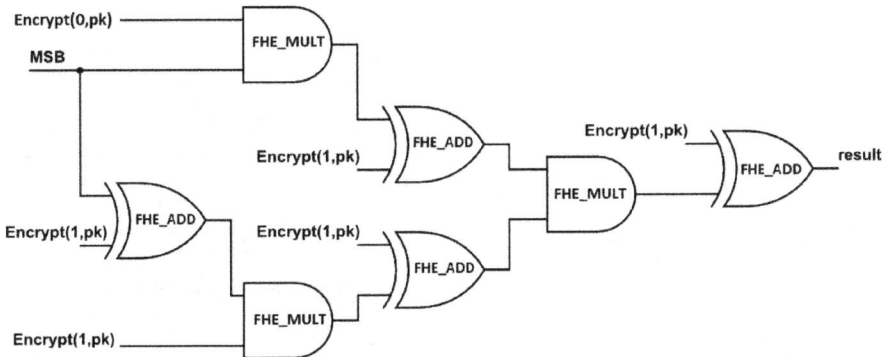

Fig. 4.2 Fully homomorphic greater comparison (*FHE_GrtEq*) (Chatterjee and SenGupta 2018)

otherwise. The detailed design of the encrypted multiplexer will be explained in Sect. 4.5.2.

3. However, this condition only checks if the MSB of the subtraction result is $Enc(0)$ or $Enc(1)$. However, if two operands are exactly equal, then the MSB is also equal to $Enc(0)$. Hence, the *FHE_GrtEq* module checks greater than equal to relation (and not exact greater than relation) between two operands.

4. Similar module *FHE_LessEq* is designed to check lesser than equal to relationship. The only difference is inverse of MSB is fed to the multiplexer as input instead of MSB.

Now, we focus on greater than (or lesser than) operation and the equality operation among two operands.

FHE Equality Check Module

FHE_Equal module is designed to check if two FHE data are equal. If two operands are equal, then FHE subtraction (output of *FHE_Sub* module) of these two operands should result to $Enc(0)$. Following steps are performed to check whether the subtraction result is 0 (Fig. 4.3):

- Each of the bits of the subtraction result is inverted (using *FHE_Inv* module).
- Bitwise multiplication is performed on resultant inverted bits using *FHE_MUL* module.
- If the subtraction result is nonzero (implies that two inputs are not equal), then at least one bit of the subtraction result must be nonzero. Hence, the multiplication result becomes zero due to the presence of inversion of the nonzero bit (or bits).

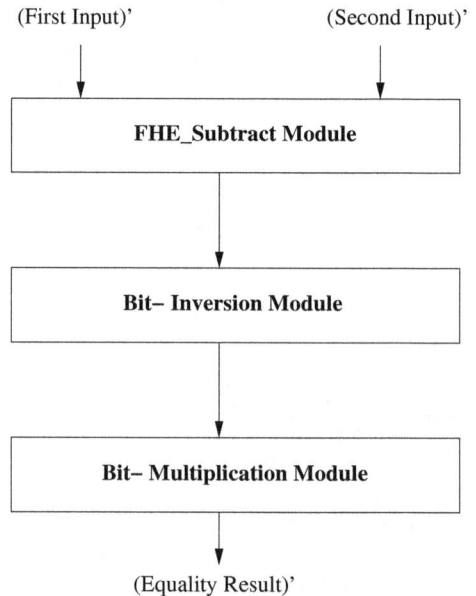

Fig. 4.3 Fully homomorphic equality check (Chatterjee and SenGupta 2018)

- Thus, *FHE_Equal* module outputs Enc(1) if two FHE inputs are equal, else Enc(0) if the two are unequal.

4.5.2 Encrypted Less Than/Greater Than Operator (*FHE_Grt* and *FHE_Less*)

In the preceding section, we discussed the implementation of greater than/ lesser than equal to operation in *FHE_GrtEq* and *FHE_LessEq* modules. In this section, we check exact greater relation between two operands by *FHE_Grt* module. Here, two same inputs a' and b' are sent to *FHE_GrtEq* module and *FHE_Equal* module. Then, the output of *FHE_GrtEq* module is multiplied (FHE_AND operation) with the inverted output of *FHE_Equal* module and the final output of *FHE_Grt* is produced. That implies if two operands a' and b' are not equal then the *FHE_Equal* module will give the output as $Enc(0)$ and hence the inversion is $Enc(1)$, then the final output of *FHE_Grt* module will depend only on the greater relationship of two operands. Similarly, *FHE_Less* module will check if a' is lesser than b' and *FHE_LessEq* is used in this case.

So far we have discussed the implementation of different encrypted operators which are important for executing the instructions of the encrypted algorithm. As shown in Fig. 4.1, these instructions can be executed only once or multiple times depending on some condition. Hence, next step of any algorithm handling is condition checking or decision making. Decision making requires one or more instructions to be evaluated and based on the evaluation result (condition), a statement or set of statements will be executed (if the condition is determined to be true). Optionally, other statements are executed if the condition is determined to be false. In the next section, we shall analyze what are the issues of decision making when data are encrypted and hence the conditions are also enciphered.

Encrypted Decision Making

Decision making in programming depends on the evaluation results of certain conditions. Depending on the conditions, certain set of instructions are evaluated, else some other instructions are executed. However, the main challenge of decision making with encrypted data is that all the conditions are checked based on the results of some encrypted operations. Since, the conditions are always in their encrypted forms, conventional equality checks to true/false values will not work. Hence, here we need an alternate strategy.

Here we shall discuss how encrypted decision module can be realized by an encrypted multiplexer (*FHE_Mux*). Multiplexer decides the output from the inputs based on the selection line. Considering the encrypted multiplexer takes two encrypted inputs A' and B', and a selector input S' (which is encrypted form of logic 1 or logic 0), the output Z' of *FHE_Mux* can be expressed as:

$$Z' = A'.\overline{S'} + B'.S' \tag{4.1}$$

Hence, considering S' as the encrypted condition, input A' will be assigned to Z', if S' is true ($Enc(1)$) else B' will be assigned to Z'. Now, consider a conditional statement where decision making is performed based on a condition, whose encrypted value is S'. Thus, if the condition is true, S' is $Enc(1)$, else $Enc(0)$. Hence, the decision making can be realized by the multiplexer circuit. Note that, such an encrypted decision making can only be realized by a multiplexer circuit (as multiplexer is a universal circuit component), which requires the underlying homomorphism to support both multiplication and addition. This again emphasizes the requirement of FHE as underlying encryption scheme.

Till now, we have discussed how to execute a code on conditional basis. Finally, we shall investigate how to execute an instruction (or set of instructions) several times that is what are the challenges for loop handling on encrypted data.

4.6 Loop Handling on Encrypted Operations

Loop allows to execute a statement or group of statements multiple times. Problem of loop handling on encrypted operations is again very similar to condition checking problem of encrypted decision making. Here are the main issues as shown in Fig. 4.4.

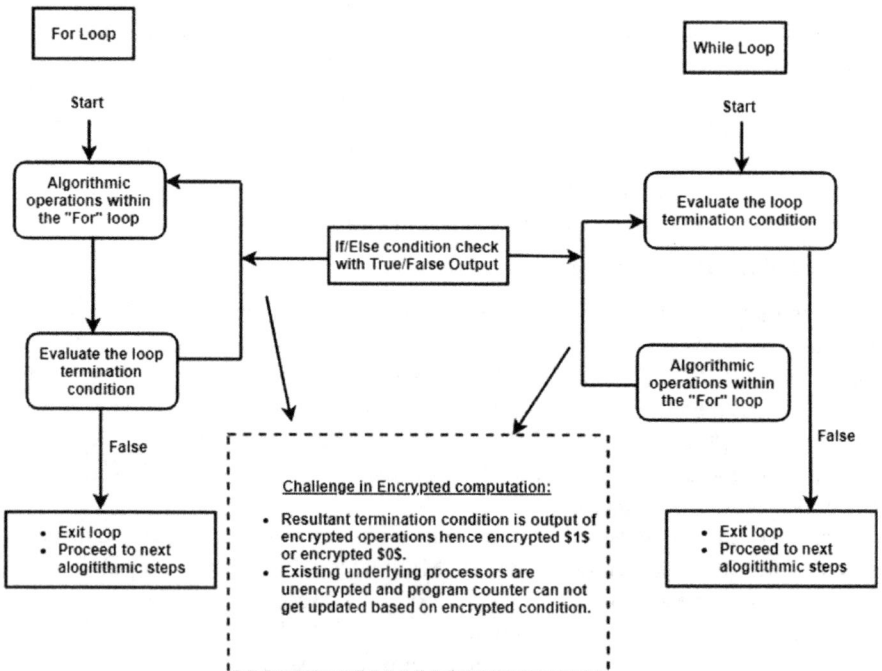

Fig. 4.4 Encrypted loop handling

Algorithm 5: Fibonacci algorithm

Input: Integer n
Output: Integer b
if $n == 0$ **then**
$\quad |\quad$ **return** 0;
else
$\quad |\quad a \leftarrow 1;$
$\quad |\quad b \leftarrow 1;$
$\quad |\quad$ **for** $i \leftarrow 3$ ***to*** n **do**
$\quad |\quad\quad |\quad c \leftarrow a + b;$
$\quad |\quad\quad |\quad a \leftarrow b;$
$\quad |\quad\quad |\quad b \leftarrow c;$

return b;

- Encrypted loop implies the execution of single or sequence of encrypted instructions. However, the initialization (or termination) condition of such loop execution is again the result of some encrypted operations and hence it is difficult to handle such encrypted conditions with present unencrypted processors.
- Traditional loop control statements cannot be handled easily. For example, we consider the *go to* loop control statement, which transfers control to a label. However, these labels should also be encrypted to be handled based on the encrypted decisions. However, such labels are not encrypted in present unencrypted processors, as the addresses are unencrypted.

Hence, terminating a loop is a major challenge while working with encrypted operations. One of the solutions could be to assume that the worst case length of the loop is utilized to terminate the loop. Further, the maximum loop length can be conceptually divided into two parts: *effective part* and *ineffective part*. In the *effective part*, the loop condition is valid, and thus the instructions executed affect the output variables. However, in the ineffective part the instructions are still executed, but they do not affect the output. In an unencrypted version of loop, the loop simply terminates when the condition becomes false, but in the encrypted loop the determination of the termination condition is not possible generally as discussed, hence redundant operations are required. However, the encrypted termination condition is determined by encrypted decision making using FHE_Mux and that further determines the length of effective part.

So far we have investigated handling the basic operators, decision making and loop handling required for implementing algorithms on encrypted data. Let us discuss a case study of traditional Fibonacci algorithm (unencrypted) and show how to translate this algorithm to its encrypted counterpart and explain the encrypted loop handling more clearly. In general, Fibonacci operation is computed with the following recursion:

with $Fib(0) = 1$ and $Fib(1) = 1$, where $Fib(n)$ denotes the Fibonacci value of variable n.

Table 4.2 Mapping of Variables (Chatterjee and SenGupta 2018)

Variables	Encrypted variables
n	`enc_n`
a	`enc_a`
b	`enc_b`
c	`enc_c`
Final fibonacci value	`fibvalue`

$$Fib(n) = 1, \quad if \ n = 0 \ or \ n = 1$$
$$= Fib(n-1) + Fib(n-2) \qquad (4.2)$$

However, handling recursion is complex with encrypted operations since existing recursion stacks of the underlying processors are unencrypted. Hence, we shall work with the non-recursive form of Fibonacci algorithm as shown in Algorithm 3.

This example is to illustrate the use of relational operators, conditional checks and loop handling when the data are encrypted. We attempt to emphasize the translation of the algorithm which operates on unencrypted data to its counterpart operating on encrypted data. Here are the different encrypted operations to execute the algorithm:

- Library libScarab (2011) maps the integer input variables to mpz_t type (shown in Table 4.2).
- For computing Fibonacci series, initially we need to compare if n equals to 0 and Fibonacci value is set to 0. With encrypted operations, this *if condition checking* is done by the following code-snippet:

```
/******* Storing all zero *********/
 for(j=0;j<len;j++){
   FHE_Encrypt(c0, pk, 0);
   mpz_set(tempZero[j],c0);
   }
 FHE_isEqual(n_comp,enc_n, tempZero, pk);
 FHE_Mux(fibvalue,fib_temp,tempZero, n_comp, pk);
```

Initially, $Enc(0)$ is stored in all encrypted bits of `tempZero` array with size `len` of 32-bit. Further, `tempZero` is compared with encrypted n value `enc_n` with FHE_isequal module. If `enc_n` equals to `tempZero`, then `n_comp` is high and it inturn selects `tempZero` as the Fibonacci value (otherwise the computed Fibonacci value for non-Zero n is stored in encrypted variable `fib_temp`). Final Fibonacci value is stored in `fibvalue`.

- Fibonacci value of nonzero numbers are computed with the next for loop as shown in the Algorithm 3, where addition between two encrypted values are computed using the mathematical addition operator as explained in Sect. 4.4.1 and then the values are assigned to other encrypted variables.

- However, the most complex part of handling this algorithm is to translate the loop iteration `for(inti = 3; i <= n; i + +)` to operate in the encrypted domain. Since, the value of n is encrypted, so is the loop termination condition. Two possible scenarios can occur in such case:

1. If the unencrypted value of the number (of which Fibonacci value to be computed) is known, then only the for loop can be iterated with unencrypted values. The following code snippet is explaining this scenario:

```
/********** For n known ******/
for(icnt =3; icnt <= fibdata; icnt++)
{
 mpz_arrayadd(enc_c, enc_a, enc_b, pk);
 for(j=0;j<len;j++)
 {
  mpz_set(enc_a[j], enc_b[j]);
  mpz_set(enc_b[j],enc_c[j]);
   }
}
FHE_mux(fibvalue,enc_b, tempZero, n_comp, pk);
return (fibvalue);
```

The variables a and b are encrypted to `enc_a` and `enc_b` with values $Enc(1)$ initially. Values of `enc_a` and `enc_b` are added using `mpz_arrayadd` module (FHE addition between two encrypted variables) and the value is stored to `enc_c`. The loop iterates till the loopcount equals to n and finally the Fibonacci value (encrypted) is stored in `fibvalue`.

2. However, the above scenario is rather impractical. Since, it is beneficial to compute Fibonacci value with unencrypted algorithms if exact value of n is already known. Otherwise if n is encrypted, the loop count variable `icnt` should be compared with encrypted n and hence both need to be encrypted as comparison is not possible with one encrypted and other unencrypted value. Finally, the comparison result is encrypted and incomprehensible by the underlying unencrypted processor. Hence, the main challenge while working with encrypted data is to identify the termination condition of the loop.

One possible solution in this scenario is that the cloud administrator may know the maximum possible value of n (`maxfibdata`) and uses that knowledge to run the code on encrypted data as shown below:

```
/********** For n encrypted ******/
for(icnt =3; icnt <= maxfibdata; icnt++ )
{
    FHE_encrypt(enc_i, icnt, pk);
    FHE_isgreater(i_comp, enc_n, enc_i, pk);
    mpz_arrayadd(enc_c, enc_a, enc_b, pk);
```

```
for(j=0;j<len;j++){
    mpz_set(enc_anew[j], enc_b[j]);
    mpz_set(enc_bnew[j], enc_c[j]);
    }
FHE_mux(enc_atemp,enc_a, enc_anew, i_comp, pk);
FHE_mux(enc_btemp, enc_b, enc_bnew, i_comp, pk);
    for(j=0;j<len;j++){
        mpz_set(enc_a[j],enc_atemp[j]);
        mpz_set(enc_b[j],enc_btemp[j]);
        }
    }
FHE_mux(fibvalue,enc_b, tempZero, n_comp, pk);
return (fibvalue);
```

In this code-snippet as discussed before, the for-loop iterates for icnt <= maxfibdata and a comparison is done to check if enc_i (encryption of i_cnt) is less than enc_n by FHE comparison module. If enc_i reaches enc_n, then the loop reaches the ineffective part and enc_a and enc_b values do not update any more. Hence, the final fibvalue equals the encrypted Fibonacci value of n, whatever be the iteration value of maxfibdata. However, cloud administrator may not always have the information about possible maximum value of the encrypted data. Further, this method in spite of being functionally correct incurs some redundant operations and increases the timing requirement. The above discussions and the case study show that detection of termination is a problem while running algorithms which operate on encrypted data. In the next section, a solution is proposed by performing a server- client message passing to solve the problem. In such a case, server sends an encrypted signal to the client, which generates an interrupt in the client side when the program need to be terminated. Client sends back another signal to the server to terminate the program. However, care has to be taken that the message exchanged does not reveal any critical information.

4.7 Encrypted Program Termination Using Interrupt

In the previous sections, we have discussed it is actually feasible to realize arbitrary algorithms on encrypted data using FHE operations. All these algorithms work based on encrypted instructions and generate the final results in encrypted form, which can be directly stored in cloud server. However, real world works on unencrypted data, hence authorized client may decrypt it finally at the client side if the unencrypted value is required for further processing. Here we discuss how to use this decryption powers to solve the problem of encrypted loop termination without leaking any critical information.

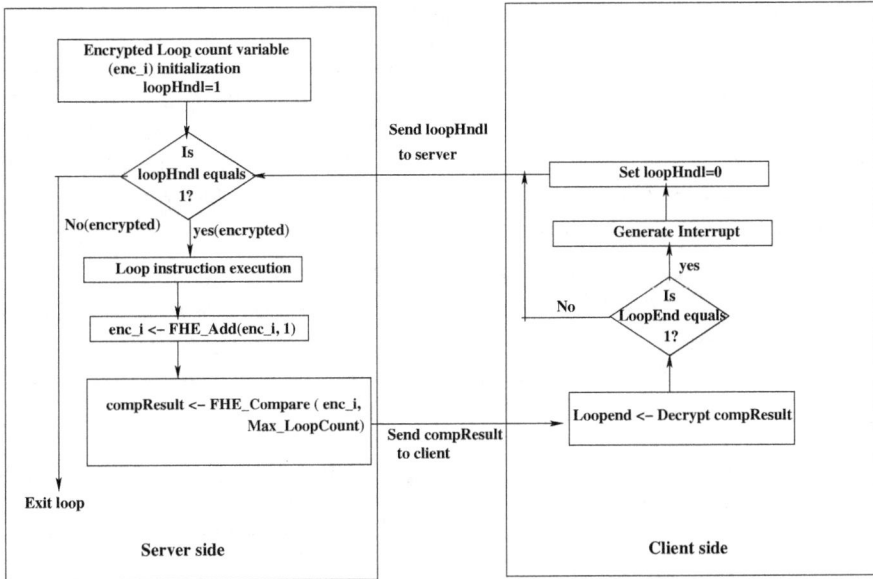

Fig. 4.5 Example of encrypted loop handling (Chatterjee and SenGupta 2018)

In Fig. 4.5, a generalized encrypted loop execution is explain with client intervention. Here, we define an unencrypted variable $loopHndl$ at the server side such that loop is getting executed if $loopHndl = 1$. Now $compResult$, the encrypted result between encrypted loop counter variable (enc_i) and encrypted value of maximum loop count) is sent to client in each iteration. For Fibonacci computation encrypted value of input data n (enc_n) is the value of encrypted maximum loop count. The $compResult$ value is decrypted at the client side and if it is 1 (indicates enc_i reached to maximum value and the loop should be terminated) an interrupt is generated in the client side to set the value of $loopHndl$ to 0. This value is sent back to server and based on this value the control exits from the loop. Since, $loopHndl$ is not directly related to the critical information of the instruction executed in the loop, this unencrypted traffic from client to server does not reveal any sensitive information. Further, server and client can settle upon some symmetric encryption key and the client can send encrypted $loopHndl$ with that key to the server. The termination of loops depend on encrypted condition checking results, they cannot be detected without the knowledge of the secret key. The same method of encrypted loop handling by server-client interaction can be used to detect program termination. However, there are some other practical challenges which will be discussed and addressed in next chapter. Up to now, we have discussed basic encrypted operators as well as decision making and loop handling on them. However, for coding any algorithm, proper organization of data is very important. Hence, choice of data structure plays a key role and different data structures are suited to different kinds of applications. In the next section, we shall discuss about defining data structures for encrypted data.

Data Structures and Encrypted Programming Approaches

Data structure is a closely related issue while considering algorithms. It is the way of organizing data items considering their relationship to each other and algorithm is the sequence of instructions to manipulate those data. Proper choice of data structures also make crucial difference in the performance of a program. Hence, this should be separately discussed while working with encrypted data. Here we find how few basic data structures can be defined on encrypted data:

- An *encrypted array* can be defined with a typecast of encrypted data type. For example, an array *int arr[n]* will be translated to *mpz_t enc_arr[n]*, where *n* is the number of elements in array and array elements are of *mpz_t* type. However, the indices of the array should be unencrypted since existing unencrypted processors can not handle encrypted array indices.

```
Struct student{                 Struct enc_item{
       int ID;       ----->        mpz_t enc_ID;
       int marks;                  mpz_t enc_marks;
   }                            }
```

After finalizing the data structure, deciding the programming approach is the next step for converting an algorithm to an encrypted program. Iterative and recursive are the two main programming approaches. In most cases, recursion provides an elegant solution and ease to the programmer in case of design. Further, iterative solutions are usually more efficient than recursive solutions as they do not incur the overhead of multiple function calls. In the next section, we shall discuss about the complexities of applying recursion on algorithms with encrypted operations.

4.8 Recursion Handling with Encrypted Operations

Recursion is the process of repeating items in a self-similar way. In recursive programming, a function calls itself repeatedly. However, every recursion should have an exit condition, otherwise the program goes to an infinite loop. However, this exit condition is again the result of some operation and in encrypted domain this exit condition is again encrypted, since it is generated from some encrypted operations. But, the existing recursion stacks of the present processors do not understand encrypted conditions. Hence, implementation of encrypted recursion is not possible using present unencrypted processors and we follow encrypted iterative approach for most of the algorithms or make use of a specially user-defined stack. In the next section, it will be discussed how stack data-type is realized for handling encrypted data.

4.8.1 Design of Encrypted Stack

As stated above, for handling recursion, system maintains a recursion stack for storing all the intermediate recursion values. While converting any recursive function to iterative, defining an auxiliary stack is necessary. For handling complex encrypted operations (which results encrypted conditions), here discuss the concept of encrypted stack, which should store encrypted data and be capable of handling encrypted push pop operations. For detailed design of encrypted stack, we refer to Sect. 3.2.3.

So far we have described how to handle an encrypted stack with encrypted operations. Queue and linked list are the two other important abstract data types (ADT). Next, we investigate the challenges of implementing these data types on encrypted data.

4.9 Design of Encrypted Queue

In the ADT queue, the entities in the collection are kept in order. The basic difference of stack with queue is that stack elements are LIFO (Last In First Out) based, whereas queue elements are FIFO (First In First Out) based. Hence, queues unlike stack requires two pointers rear (the end where data is inserted or enqueued) and front (from where data is extracted or dequeued). However, front and rear should also be encrypted, since we are discussing about encrypted queue. Further encrypted enqueue is very similar operation with pushing data into the encrypted stack, similarly encrypted dequeue operation resembles pop operation from encrypted stack. However, in this case enqueue operation modifies the encrypted rear value, while the deque operation modifies the encrypted front value.

4.10 Design of Encrypted Linked List

Linked list is a dynamic data structure which consists of group of nodes and each node contains data as well as link address of the next node. For unencrypted domain, data as well as the address both are unencrypted. However, designing linked list with encrypted data raises two options:

- In the first option, data is encrypted but the address remains unencrypted. In this situation, the memory location can directly be accessed by the reference field of the node. Encrypted operations will be required only to process data and not the addresses.
- However, situation may appear that both data and link address need to be encrypted and the list traversal becomes more complex since the present underlying processors contain unencrypted address space.

It is well known that the structure representation of a linked list with unencrypted data is as follows:

```
struct node {
  int x;
  struct node *next;
};
```

To translate this list to an encrypted one of the first type mentioned above, we can first consider a structure with encrypted data and unencrypted addresses. Thus:

```
struct enc_node {
   mpz_t enc_x;
   struct enc_node *next;
};
```

However, second type of linked list can not be realized in the same way since the link should also be encrypted. Hence, a separate encrypted address field is defined to store the encrypted link. Such a list node is hence represented as:

```
struct enc_node {
   mpz_t enc_x;
   mpz_t enc_address;
};
```

Traversal of first type of linked list (shown in Fig. 4.6 as L1) is almost similar to unencrypted linked list, since the address field is not encrypted. Homomorphic operations are only required to manipulate the list data. Now, we shall concentrate on the second type of linked list (shown in Fig. 4.6 as L2), which is more complex one (with both encrypted data and encrypted link address) and investigate how the link traversal, data can be performed. In the list L1 and L2, each node contains data d_i and link to the next node (denoted by M_i or $Enc(M_i)$).

Traversal of Encrypted Linked List with Encrypted Address

Traversing linked list with encrypted addresses requires identifying the actual address (unencrypted) location from the encrypted address, since data actually resides in unencrypted memory locations in current processors. In this case, the existing encrypted link addresses of the list nodes are compared with encrypted values of each memory location (using *FHE_Equal* module). Depending on the comparison value (if a match is found) for any encrypted address, the corresponding unencrypted address will be considered as the next reference address of the node.

However, this way of traversing linked list is not very practical. Since for every location of memory an encryption and then comparison need to be performed with the encrypted link, hence the time requirement is too high considering a moderate

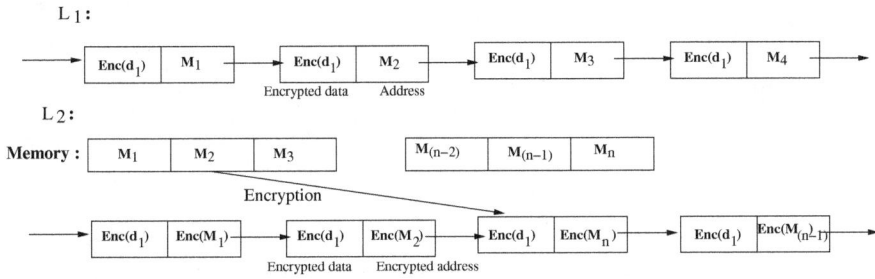

Fig. 4.6 Encrypted linked list (Chatterjee and SenGupta 2018)

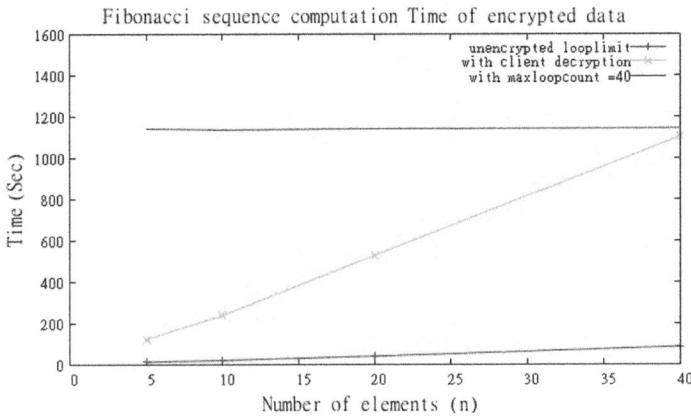

Fig. 4.7 Fibonacci time on encrypted data

memory size. Due to the complexity of traversal, adding or removing data from a linked list (encrypted) is also time consuming since it again requires encryption of each memory location and further comparison with encrypted references in each node.

Timing Analysis of Different Encrypted Operations

Figure 4.7 shows the timing requirements of different approaches to compute Fibonacci sequence on encrypted data. It shows that time requirement increases linearly with the increase of n. However, if the maximum loop count is fixed to a large value of n, the timing is independent of chosen n, for which Fibonacci sequence is being computed.

Table 4.3 shows required time to execute different operations on encrypted FHE data (each data is the encryption of 32-bit integers) and gives an idea about times of executing encrypted algorithms using these operations. Table 4.4 shows timing requirement of few other traditional algorithms implemented using FHE operations. All these examples are implemented using client-server interaction for loop handling and program termination and evaluated for correctness on a Linux Ubuntu 64-bit

Table 4.3 Timing requirement of encrypted operations (Chatterjee and SenGupta 2018)

Operations	Time (s)
	Encrypted domain
Arithmetic operators	
Addition	1.64
Subtraction	3.28
Multiplication	128
Division	600
Relational operators	
Less than	19.2
Greater than	19.2
Equal to	4.9
Less than equal to (leq)	14.2
Greater than equal to (geq)	14.2
Condition check	
If-else	5.8

Table 4.4 Timing requirement of encrypted operations (Chatterjee and SenGupta 2018)

Encrypted operations	Time (s)
Average of 100 numbers	764
Binary search within 100 numbers	3271.6
Bubble sort with 100 numbers	13580

machine with $i686$ architecture 1.6 GHz processor. Constant researches are going on to improve the performance of FHE encryption scheme. Here timing requirement has been shown considering underlying Library libScarab (2011) based on the work (Perl et al. 2011). However, the basic FHE addition and multiplication modules can be replaced by some upcoming faster implementations in future to improve the performance, otherwise the overall implementations and methods of encrypted operations handling will remain same as proposed in this chapter.

Further, some encrypted abstract data types (encrypted stack, encrypted queue and encrypted linked list) have been proposed. Table 4.5 shows the required time for executing encrypted operations on these abstract data types. Further, encrypted linked list traversal requires (encryption time + comparison time) for each memory address as explained in Sect. 4.10. For fixed block size, this approach becomes gradually costlier with the increase of memory size. On the other hand, working with already encrypted memory as an underlying architecture is difficult, since our existing processors are unencrypted. This is one of the motivations to design encrypted memory, however this will be illustrated in the next chapter.

Table 4.5 Timing
requirement of encrypted
ADT operations (Chatterjee
and SenGupta 2018)

Encrypted operations	Time (s)
Stack operations	
Push	7.5
Pop	9.1
Queue operations	
Enqueue	8.2
Dequeue	10.6

Application of the Proposed Methodology for Cloud Computing

The above methods for translating algorithms to operate on encrypted data is proposed as a methodology for assisting existing solutions for outsourcing computations on the cloud. Two viable methods proposed are mentioned below to pinpoint
the utility of the proposed solution. One of the technology is **Token based Cloud
Computing**, that uses the concepts of secured hardware tokens (reza Sadeghi et al.
2010). The other one is the **Twin Clouds**.

- **Token based Cloud Computing** is tamper-proof against physical attacks but can
  perform arbitrary computations to enable the cloud client to perform confidential
  computations on the cloud side. However, if the whole security critical computation
  takes place in the token, the operation is not really taking place in the cloud (reza
  Sadeghi et al. 2010). There would be situations where a fast, timely and yet private
  computation is required, like medical records. In this scenario the token based cloud
  computing paradigm computes with the client (much like the trusted cloud), and
  outsources the arbitrary computations to a storage cloud. Storage cloud performs
  computations of secret circuits using FHE, where again proposed techniques can
  be useful.
- **Twin Clouds** (Bugiel et al. 2011): In this architecture, two clouds (a trusted
  cloud and a commodity or public cloud) are considered. The client connects via
  a secured communication channel (like SSL, TSL) with the private cloud, which
  encrypts the database and outsources the data and computations to the public
  commodity cloud. The outsourced data to the public cloud must be confidentiality
  and integrity protected and should support arbitrary computations. Mentioned
  solutions to translate an arbitrary algorithm to operate on FHE data provides the
  first steps to make architectures like Twin Cloud feasible.

4.11 Conclusion

In this chapter, it is highlighted that although FHE provides the capability of performing computations over encrypted data, it leaves several challenges for executing
algorithms which run over encrypted data. In this pursuit, discussions involve the

issues in translating the variable definitions, instruction executions, handling of loops and terminating conditions when the algorithms handle encrypted data and encrypted controls. Through examples and actual implementations, the complexities of handling encrypted arrays, modifying data-structures like stacks, etc are illustrated in particular, challenges of translating recursive codes to their encrypted counterparts. In the subsequent chapter, we focus on how to extend these encrypted operators for database operations.

References

Knuth DE (1973) The art of computer programming, volume I: fundamental algorithms, 2nd edn. Wesley, US

Chatterjee A, SenGupta I (2018) Translating algorithms to handle fully homomorphic encrypted data on the cloud. IEEE Trans Cloud Comput 287–300

Library libScarab (2011). https://github.com/hcrypt-project/libscarab

Rass S, Slamanig D (2013) Cryptography for security and privacy in cloud computing. Artech House Inc, Norwood, MA, USA

Perl H, Brenner M, Smith M (2011) Poster: an implementation of the fully homomorphic smart-vercauteren crypto-system. In: Chen Y, Danezis G, Shmatikov V (eds) ACM conference on computer and communications security, ACM, pp 837–840

reza Sadeghi A, Schneider T, Win M (2010) Token-based cloud computing secure outsourcing of data and arbitrary computations with lower latency. In: Workshop on trust in the cloud

Bugiel S, Nürnberger S, Sadeghi AR, Schneider T (2011) Twin clouds: secure cloud computing with low latency. In: Proceedings of the 12th IFIP international conference on communications and multimedia security, CMS'11, pp 32–44

Chapter 5
Secure Database Handling

Onset of cloud computing allowed various IT services to be outsourced to cloud service providers (CSP). This includes the management and storage of users' structured or unstructured data called Database as a Service (DBaaS). However, users need to trust the CSP to protect their data, which is inherent in all cloud-based services. Enterprises and Small-to-Medium Businesses (SMB) see this as a roadblock in adopting cloud services because they do not have full control of the security of the stored data on the cloud. Security may be enhanced with the use of hybrid cloud, which is a cloud computing environment combining on-premises, private cloud and third-party public cloud services. Critical applications which demand higher security may be moved to private cloud. However, that incurs added cost and does not provide full security conformation.

Few reports (Bradford 2018) have shown "Security" is a serious bottleneck while making cloud popular in actual business. In 2010, Microsoft faced a breach when non-authorized users of the cloud service got access to the employee contact due to some configuration issues of its Business Productivity Online Suite. In 2012, Dropbox reported unwanted tapping of more than 68 million user accounts and nearly 5 GB of data. In April 2016, the National Electoral Institute of Mexico experienced a breach of 93 million voter registration records. Similar hacking incidents were experienced by Linkedin, Yahoo including high profile breaches in Apple iCloud.

Due to all these reasons entrepreneurs are not convinced enough to store their critical data only trusting the security promises made by the cloud. Hence, it is important to analysis possible threats in cloud data storage.

© Springer Nature Singapore Pte Ltd. 2019
A. Chatterjee and K. M. M. Aung, *Fully Homomorphic Encryption*
in Real World Applications, Computer Architecture and Design Methodologies,
https://doi.org/10.1007/978-981-13-6393-1_5

Table 5.1 Cloud classifications

Architecture	Specification
Private cloud	Exclusive use by single organization comprising multiple user
Community cloud	Exclusive use by a specific community of users from multiple organizations that have shared concerns
Public cloud	Open use by the general public
Hybrid cloud	A combination of at least two of the above three

5.1   Security Issues in Cloud

Table 5.1 mentions major classifications of cloud platform. Data integrity, availability and data security are three major concerns while storing data in almost any type of cloud (Huang et al. 2014). In Table 5.2, major approaches of these issues are mentioned. There are following schemes in literature to ensure efficiency, unbounded use, and self-protect mechanism in data integrity:

- POR and Provable Data Possession (PDP).
- Proof of Erasability (POE), where cloud ensures comprehensive destruction of stored data in the storage when client withdraws the data and disassociate with the storage provider.
- Proofs of Secure Erasure (PoSE-s) also has similar functionality like POE with remote access support.

Availability ensures data are accessible and usable when authorized users request to access. Other than reliable hardware infrastructure. This is mostly implemented by different data backup strategies as discussed in Table 5.2. Other than these two issues, cloud data are prone to different adversarial attacks as highlighted in Table 5.3 (Grubbs et al. 2017). Due to these security issues, one possible solution for the data owners is to store their sensitive data in encrypted form. However, to take full advantage of cloud database as service, there should be a solution to process on the stored data while it is encrypted in cloud and Fully Homomorphic Encryption (FHE) is a promising solution in this aspect.

5.2   Sate of the Art

We are intended to discuss about standard query language (SQL) and discuss the challenges of realizing basic SQL queries in encrypted domain. Relational database management system (RDBMS) is one of the most popular way to perform processing on database. In this section, some introductory information about the state-of-the-art will be provided focusing on two existing encrypted databases: ZeroDB and CryptDB (Miguel et al. 2016).

Table 5.2 Issues in cloud data

Scheme	Specification
Data Integrity	
Proof of Erasability (POE)	Comprehensive destruction of stored data in the storage when clients withdraw the data and disassociate with the storage provider
Proofs of Secure Erasure (PoSE-s)	This scheme replaces hardware-based attestation with remote attestation to update secure code and secure storage erasure for cloud
Proofs of Retrievability (POR) (Juels et al. 2007)	Scheme proves stored data are intact in server during the storing and retrieving process of the client
Improved POR scheme (Lillibridge et al. 2003)	Pre-processing requirement of the (Juels et al. 2007) scheme removed by storing redundantly encoded data blocks with message authentication code (MAC)
Provable Data Possession (PDP) (Ateniese et al. 2007)	PDP is more efficient. POR ensures both data integrity and retrievability, whereas PDP guarantees only data integrity at cloud
Availability	
Data backup (Huang et al. 2014)	Amazon EC2 and S3 keeps back up in different availability zones
Incremental backup with delta encoding (Huang et al. 2014)	Exploits the correlation between current files with previous backup version and stores the differences
Data deduplication on encrypted (Storer et al. 2008)	Specialized data compression technique that identifies common data chunks across data
Predict future availability failure (Guan et al. 2011)	Using Bayesian methods and decision trees
Security	
Access control (Sahai et al. 2005; Ahn et al. 2000; Sandhu et al. 1996)	Prevents privilege Abuse/ Elevation attribute-based access control (ABAC) Role-based access control (RBAC)
Inference policy	Interpretations from certain data analysis or facts are protected at certain higher security level
User identification/authentication	Authentication keeps the sensitive data safe and from being modified by unauthorized user
Searchable encryption (Boneh et al. 2004)	Encrypt a search index generated over a collection of data in such a way that its contents are hidden without appropriate tokens, which can only be generated with a secrete key
Fully homomorphic encryption	Supports encryption of data and arbitrary operations on stored cloud data

Table 5.3 Possible attacks on cloud data storage

Attacks	Details and damage
Snapshot attacker (Grubbs et al. 2016; Kumar et al. 2011; Lewko et al. 2010)	Can only obtain a single static observation of the compromised system Damage: attacker can obtain image of the virtual machine executing the DBMS or rootkitting the OS
Persistent passive attacker (Grubbs et al. 2016)	Passively observes all database operations including the queries issued and how they access the encrypted data Damage: Damaging in encrypted databases based on property-revealing or order preserving encryption
SQL injection (Guimaraes et al. 2009)	Injection of arbitrary code Damage: Takes full control of the memory space of the DB process
Virtual machine (VM) image leaks (Balduzzi et al. 2012; Ristenpart et al. 2010)	Full-state snapshots are leaked including the VM's memory and CPU registers Damage: Attack yields the persistent and volatile OS and DB state
Full-system compromise (Verizon data breach 2016)	Rooting the DBMS Damage: gaining full access to the persistent and volatile OS and Database state

ZeroDB is an encrypted database which enables clients to perform end to end encrypted operations without exposing encryption keys or plaintext data to the database server (Egorov et al. 2016). Operations in general are search, sort, query and share. ZeroDB is based on ZODB implementation of OODBMS which uses object references instead of relational tables. The server acts as a storage system with consistency guarantees and has no insight into the type of data being stored, which eliminates data being exposed via a server side breach (only encrypted data would be visible in case of a successful infiltration). Data in ZeroDB is stored in BTrees, and encrypted records are stored as indexes (or nodes) in the BTree. BTrees are a generalization of the binary search tree; instead of storing one key and having 2 children nodes, BTree nodes have "n" keys and (n+1) children nodes. The height of the tree is smaller and requires much less disk access than a binary search tree would, optimizing read accesses that ZeroDB needs for frequent round trips during a client query. Time complexity remains O(log n) for search, insertion, and deletion. In ZeroDB, a BTree consists of encrypted buckets where each bucket is either a root, branch, or leaf node, and the leaf nodes of the tree point to the objects being stored. Performing queries is done by traversing the tree downwards to fetch records as and when you need them. Since everything is encrypted on the server side, the server does not know how the individual objects are organized in a BTree, or whether the object belongs to an index. Since ZeroDB's query protocol involves multiple round trips between the client and server, frequently accessed queries are cached on the client side to avoid unnecessary network calls. In ZeroDB's query protocol, the client makes a request to the server to return an encrypted node, decrypts it, makes a decision on which node to traverse next, and then requests the server to return the next encrypted node. The client traverses the BTree remotely until it retrieves the required record/s or object/s. However, this protocol allows all data processing to happen on the client side, in plaintext. Main limitation of ZeroDB is that Querying involves multiple query roundtrips between the client and server when the client traverses the encrypted BTree on the server.

StealthyCRM (Rodel et al. 2015) is a customer relationship management (CRM) application that uses an encrypted database management system. This database integrates CryptDB (Popa et al. 2011) with homomorphic encryption to provide secure DBAAS for CSPs. CryptDB provides confidentiality for applications supported by encrypted SQL databases against security attacks. CryptDB executes SQL queries on encrypted datasets with well-defined operators making it practical to process on encrypted data (Popa et al. 2011). CryptDB achieves this by encrypting data in different layers of encryption scheme, or onions of encryption. However, in CryptDB, types of queries are limited due to the encryption schemes used. StealthyCRM uses open-source implementation of FHE called HELib (Homomorphic Encryption Library) (HElib 2018) to integrate FHE with CryptDB. StealthyCRM enables to perform arbitrary operations on encrypted dataset. Thus, on StealthyCRM, more varieties of queries can be processed securely. In contrast with ZeroDB, all SQL operations in StealthyCRM happen on the server side (CSP) while the data is encrypted. Data will never be decrypted on the server, ensuring the privacy of the data owners. It is not possible to compute complex SQL queries (like addition followed by non-equality comparison) considering the processing capabilities of partially homomorphic encryption in CryptDB. FHE enables arbitrary computation on encrypted data by integrating CryptDB with HELib (HElib 2018) to process these complex queries. StealthyCRM framework extends the encryption algorithms of CryptDB to include leveled FHE of HELib to support homomorphic addition and multiplication. Before storing encrypted data in StealthyCRM, the system creates encrypted tables and encrypted objects on the server side. The encrypted tables are created as per CryptDB specification and the encrypted objects are created by StealthyCRM using the HELib. The StealthyCRM system forwards the queries to the CryptDB Server for simple queries and to the StealthyCRM's encrypted object handlers for more complex queries. StealthyCRM analyses the query and rewrites it into two new queries - one for CryptDB (which analyses text queries) and one for StealthyCRM's encrypted object handler. The proxy decides whether CryptDB or StealthyCRM is involved for a particular query and forwards the encrypted queries to the server side. The server sends back the encrypted result to the application after execution. StealthyCRM's decryption module decrypts the fetched result and passes it back to the application generating the query. Similar to ZeroDB in StealthyCRM, customer's data will always be encrypted at the CSP.

StealthyCRM has limitation on the number of records as well as queries that can be supported. Table 5.4 is showing performance for executing some standard TPC-C queries, where TPC-C stands for an online transactional processing system bench-

Table 5.4   The StealthyCRM performance for different types of TPCC queries (Rodel et al. 2015)

Query	MySQL (ms)	StealthyCRM (ms)
Create	10	120
Insert	1.28	50
Select sum	0.1	16.2
Select join	0.19	190
Select range	0.16	90

mark (TPC 2018). The table shows the order of performance degradation when same query is executed with encrypted computation support. In the subsequent sections, we discuss about a generalized approach of implementing such queries with primitive FHE circuits, so that these designs can be easily improved in future replacing with faster FHE addition and multiplication gates for better performance following the same design approach.

5.3   Basic Operations for Database Handling and Their Encrypted Variants

Database is organized collection of data and there are few basic operators to perform processing on the data. Since, data is FHE encrypted the database related operators should also need to be designed in special way. Following are the major encrypted operators that have been discussed in the previous chapter:

- Arithmetic operators
- Comparison operators
- Logical operators
- Bitwise operators

We shall continue our discussion to develop database related operators in encrypted domain and challenges of handling special database. Introducing relational database model in encrypted domain requires an extra effort to realize all the operators as well as database expressions with FHE operations. Further, database expressions are combination of multiple operations, results of which also should be combined with FHE operators. Creation of such database is straightforward, where FHE encrypted values are kept in respective database locations. Dropping of database removes all the entry of the database, in spite of the fact they are FHE encrypted. In the subsequent sections, we observe how to realize different SQL expressions in encrypted domain with some preexisting FHE primitives.

5.3.1 INSERT and SELECT Operation

Data storage and retrieval are two most needed operations for data management. In general, *INSERT* and *SELECT* are most known expressions used for data storage and retrieval respectively. *INSERT* operation follows the following syntax:

INSERT   INTO   TABLE_NAME   (column1,...columnN)   VALUES   (value1,... valueN);

Although we are dealing with FHE data, insertion of FHE values in the columns of the database table is quite straightforward. The column names are defined as unencrypted strings and FHE values are inserted as the column elements directly, which does not involve any specific FHE operations. Similarly, *SELECT* operation of database usually takes the following syntax:

SELECT column1, column2,...columnN FROM table_name

Selection of particular column can be done by direct string matching. If the select operation is defined as *SELECT * FROM table_name*, then all the columns are directly selected. However, these selections are not conditional, hence do not involve any FHE operations. In database management system (DBMS), *WHERE* clause is used to specify a condition while fetching data from single table or joining multiple tables. These conditions are generated from FHE operations, hence encrypted condition handling need to be incorporated in *SELECT* as well as *UPDATE* (to refresh database) statements.

5.3.1.1   Conditional Encrypted *UPDATE*

WHERE clause in database is used for extracting (for *SELECT* operation) or database overwriting of specified records whenever specific conditions are satisfied (Fig. 5.1).

Figure 5.2 shows how *WHERE* clause can be realized while implementing the database *UPDATE* statement. For the rest of the paper, we consider the encrypted database EDB with n rows and m columns. Each of the inputs of the database can be represented as Val_{ij}, for n rows of database each $1 \leq i \leq n$ and for m columns each $1 \leq j \leq m$.

As an example, let's consider EDB consists of identification number (ID), age (AGE) and salary (SAL) columns, where values are encrypted with FHE. We input two values:

- Condition input (*condInput'*) based on which FHE condition checking will be performed.
- *newlimit'*: If FHE conditions under *WHERE* clauses are satisfied, this new value will update the database field. This may be a whole row or a single field value.

Based on these assumptions *UPDATE* query is modified in the following way:

UPDATE EDB SET AGE = newlimit' WHERE SAL > condInput'

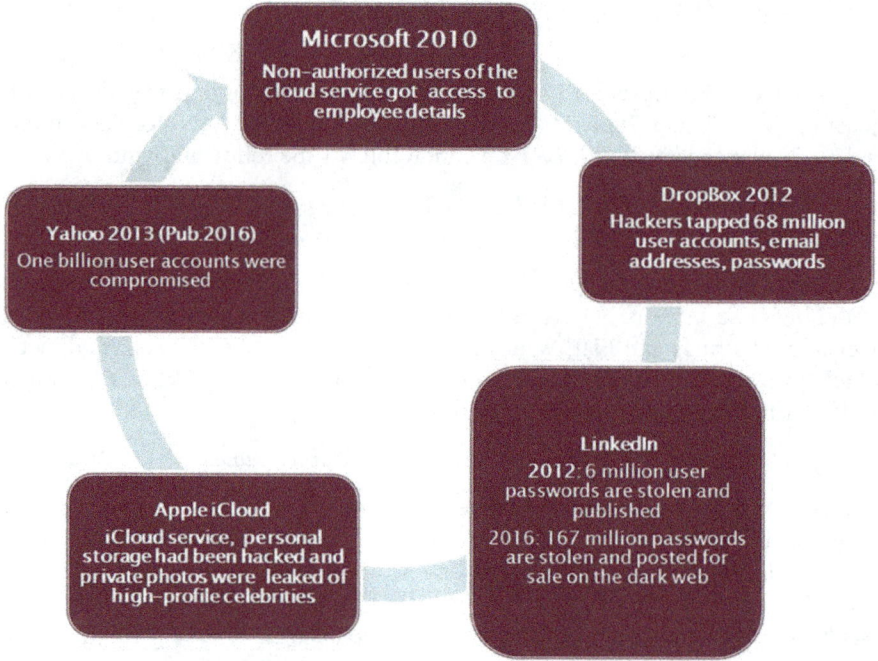

Fig. 5.1 Popular security breaches in cloud

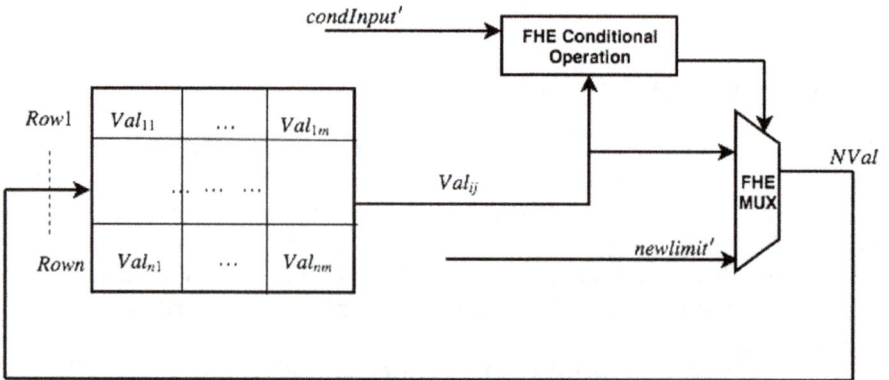

Fig. 5.2 Conditional encrypted update operation

As shown in Fig. 5.2, the update operation is implemented on comparing database inputs. The FHE Conditional Operation module can be used to implement any operations for FHE condition checking and the output of FHE condition checking, *NVal* will update the refreshed database field. In general the condition checking operations are implemented by FHE greater/lesser or equal to operators as detailed in the previous chapter.

5.3.1.2   Conditional Encrypted Select

In general, *WHERE* clause can be represented as follows with *WHERE* for conditional selection:

SELECT Database_fields FROM EDB WHERE (condition)

Implementation of *WHERE* clause with *SELECT* operation poses extra challenge in terms of selecting the rows of the database, those are satisfying the specified condition. Before describing the *SELECT* query implementation, we define the NULL Constraint considered to design the database. We set up the initial constraint that none of the fields will have null as valid value. That sets $Val_{ij} \neq Enc(0)$ as an initial condition.

Figure 5.3 shows conditional select operation, where based on some FHE conditions (performed by FHE Conditional Operation module) database rows (all or partial values from single row) are selected. Implementation of condition checking part is similar as explained in Sect. 5.3.1.1. Encrypted *AND* and *OR* operators are used to combine results out of multiple encrypted conditions. Here, we explain the approach with a simple example.

In our previous example database EDB, conditional select can be represented as:

SELECT ID FROM EDB WHERE [condition1] AND/OR [condition2]...AND/OR [conditionN];

For simplicity, we are restricting ourselves upto two conditions, where $condition1$ is SAL $geq\ conInput1'$ and $condition2$ is AGE $geq\ conInput2'$. The conditions $conInput1'$ and $conInput2'$ can be implemented with general FHE operators. Finally, AND and OR operations are used to combine these multiple conditions. Let the output after performing FHE based comparisons be $cond'$, which will be either $Enc(0)$ (if selection condition is not satisfied) else $Enc(1)$ (if selection condition is satisfied). Then, $cond'$ should be multiplied with $InputVal'$ (ID column values from each row in this example) to get the final resultant values.

Main bottleneck in encrypted select operation is that if $cond = Enc(0)$ (selection condition is not satisfied), in encrypted domain that selects an encrypted row filled with default (or $Enc(0)$) values. In our onward discussion, we shall mention such rows as *invalid rows*. In the next section, we shall discuss what are the challenges to remove these *invalid rows* while fetching the final selected valid rows to evaluate the SQL *TOP* expression in encrypted domain.

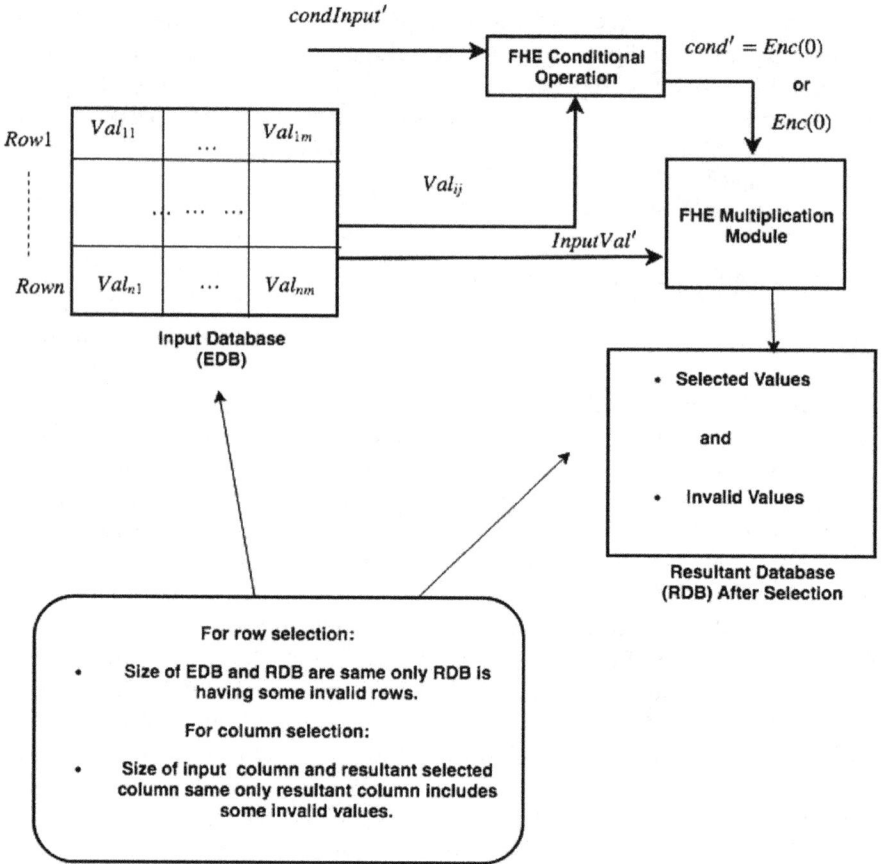

Fig. 5.3 Conditional encrypted select operation

5.3.2   TOP

As discussed, the conditional select operation is performed in encrypted domain, where removal of invalid values is not possible by bitwise matching with $Enc(0)$ since the values are generated from underlying randomized encryption. If such invalid rows can be eliminated by some unencrypted condition checking, that will infer the underlying cryptosystem is succeptible to CPA (Refer to the details of CPA in Chap. 2). One practical option to remove the invalid rows/values is to send the whole resultant dataset to the client side and the required removal can be done after decryption using the client's decryption key. However, the major drawback of this method is huge data transfer (number of rows in resultant dataset is equal to the number of rows in actual database) from server to client.

Figure 5.4 shows the implementation of *TOP* clause to fetch TOP **N** number of records from a table. In case of encrypted database handling, this information **N**

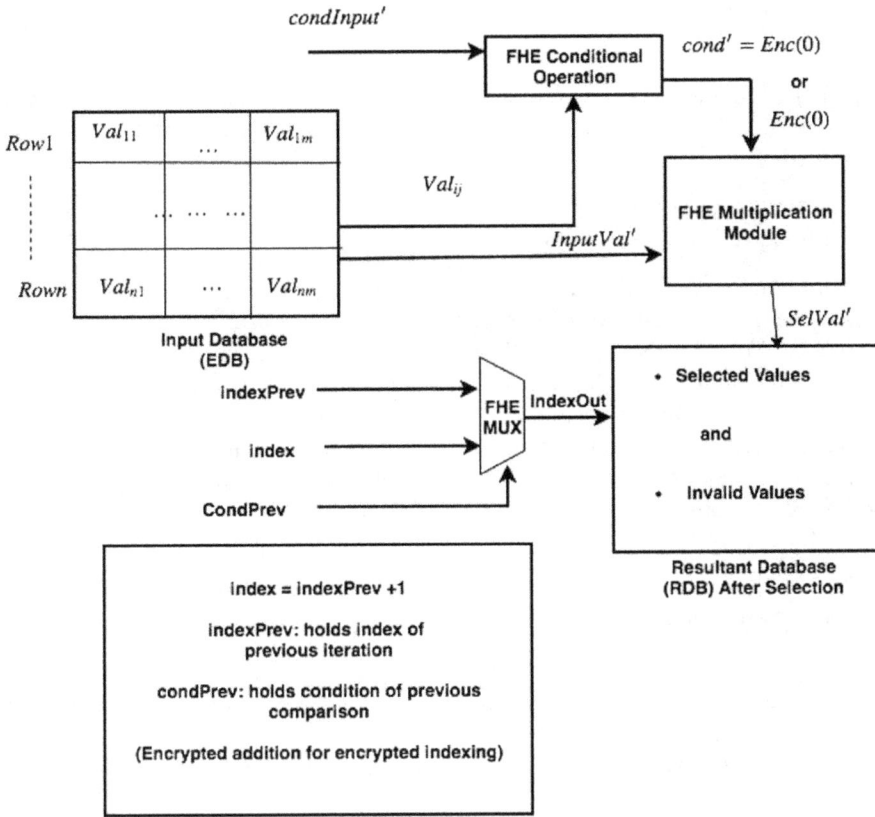

Fig. 5.4 Encrypted Top N values selection

can be provided in unencrypted form. General encrypted SELECT query will be executed on the database columns and an unencrypted for loop bounded by value N on top of it performs *Top* clause. However, as discussed in the previous section results of encrypted conditional select include invalid rows. Hence, selection of *TOP* N rows is not just the straightforward selection of top N rows from the resultant. Here we discuss an array based technique of clubbing all the nonzero values and selecting top N valid values from them.

For that reason, we store the condition of each state as *condprev* and at each state we check the value of this previous state condition. $condprev = encrypt(0)$ indicates the last appended row in the resultant column is an *invalid row*. Hence, the next result from the select operation will be appended on the same position and index of the last row will not be updated if the last row or value selection is invalid. For a database of R number of rows, this in turn arranges N valid rows on top of (R-N) invalid rows.

5.3.3 LIKE

SQL LIKE operator is for pattern matching in conditional SELECT and UPDATE operators along with WHERE clause. In SQL two types of pattern matching is mostly supported:

- The percentage sign (%) that represents multiple characters.
- The underscore (_) represents single character.

While implementing the LIKE operator on encrypted data, the comparison need to be done in encrypted way with respect to the corresponding ASCII values to find the match. For example, if any SQL query tries to execute with *SELECT * FROM DATABASE where columnN LIKE a%*, that indicates it is required to find any values that start with "*a*" in the columnN. To implement the LIKE operator in encrypted domain, it is required to compare all the columnN values to find which all starts with *a*. This comparisons will be based on ASCII values of the corresponding characters present at a specified location of the data element.

5.3.4 ORDER BY

In database, the *ORDER BY* keyword is used to sort the result-set in ascending or descending order. General syntax of *ORDER BY* is:

SELECT column1, column2, … FROM table_name ORDER BY column1, column2, … ASC | DESC;

Details of encrypted sorting are discussed in the previous chapter. While implementing the *ORDER BY* operation, simplest sorting techniques can be chosen that permits a rather straightforward implementation using only primitive comparison and swap operations (Chatterjee and SenGupta 2013). The Columns with integer values can be directly sorted, however columns with strings/character values need to be sorted based on the respective ASCII values.

5.3.5 GROUP BY

Similar to *ORDER BY*, *GROUP BY* is another operation which is used to arrange identical data into groups with some aggregate functions (COUNT, MAX, MIN, SUM, AVG) applied to group. The basic syntax of *GROUP BY* clause is as follows:

SELECT column1, column2, …, aggregate_function (expression) FROM table_name [WHERE condition] GROUP BY column1;

Figure 5.5 shows basic difference between *ORDER BY* and *GROUP BY* operations. *ORDER BY* operation performs only sort on the database columns. However, *GROUP BY* operation allows added aggregate functions on the sorted group.

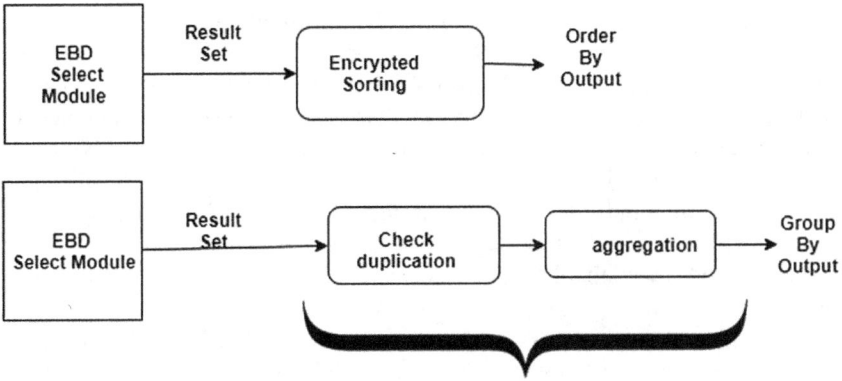

Fig. 5.5 ORDER BY and GROUP BY Operation

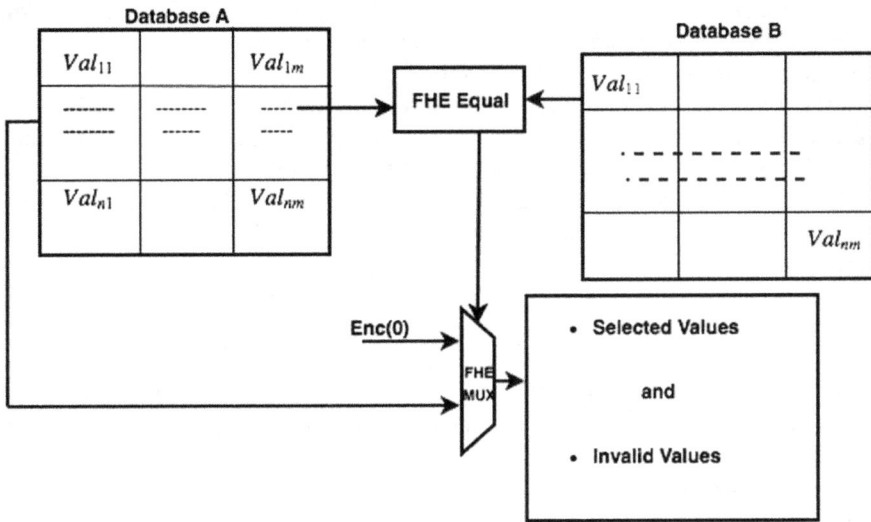

Fig. 5.6 Encrypted database join

5.4 Advanced SQL: Encrypted JOIN

JOIN clause combines rows from two or more tables, based on a related columns between them. Actual implementation is explained with the following example:

Let us consider two tables Orders (Database A) and Customers (Database B) with the related column CustomerID. Figure 5.6 shows the execution of the following JOIN command between tables Orders and Customers:

SELECT Orders.OrderID, Customers.CustomerName, Orders.OrderDate FROM Orders INNER JOIN Customers ON Orders.CustomerID=Customers.CustomerID;

Since, all the values of the tables are FHE encrypted, an FHE_Equal module checks the equality of the CustomerID value in both the tables. Design of FHE_Equal module is executed performing FHE subtraction operation between the inputs from both the tables followed by decision making based on the most significant bit (MSB) of the subtraction result. Detailed design of FHE_Equal module can be found in previous chapter following the work in Chatterjee and SenGupta (2018). Finally, if a match is found (that indicates FHE_Equal module outputs $Enc(1)$), rows from the two tables will be appended. One major challenge of implementing the "if" condition is that an invalid row with all $Enc(0)$ value will be appended to the resultant joined table.

For the sake of simplicity, we have only considered (INNER) JOIN, that returns records with matching values in both tables. Implementation of LEFT (OUTER) JOIN (returns all records from the left table, and the matched records from the right table), RIGHT (OUTER) JOIN (return all records from the right table, and the matched records from the left table) and the FULL (OUTER) JOIN (returns all the records when there is a match in either left or right table) can be implemented with few minor changes in this implementation.

Similar SQL expression $UNION$ combines results of two queries (generally $Select$ statements) into a single table without returning any duplicate rows. Hence, implementation of $UNION$ requires the usage of FHE_Select module as well as selection of distinct rows. Each distinct value selection requires traversal of the entire columN, encrypted comparison as well as encrypted decision making. Hence, that leads to large number of encrypted operations and high performance overhead of $O(n*m)$ complexity, where n and m are number of rows in input databases. In encrypted domain, this performance overhead becomes severe as the the number of rows in resultant database after select operation are same as the original database due to the presence of invalid rows. Due to this huge performance requirement, we are skipping the discussion about of implementation of union operation, which is expected to be too impractical.

When discussing about secure database handling, SQL injection is one of the topic need to be considered. In the next section, we shall investigate if SQL injection is really feasible when database is FHE encrypted.

5.5 SQL Injection on Encrypted Database

SQL injection is a code injection technique in which illegitimate SQL statements are inserted into an entry field for execution, that leads to attack or leak information from database. For unencrypted databases, an example will explain the SQL injection with the following query:

$$SELECT\ ^*\ FROM\ Users\ WHERE\ USERID\ =\ User\ Input$$

Let the field USERID should be provided by user while querying an unencrypted database Users. If there is no measure to check invalid inputs to the mentioned field, inputs like User Input = "1 = 1" may extract all the columns of the database. Any hacker can take the advantage of this query to extract sensitive information from database like UserID or Password. Now, the question arises whether such type of SQL injection scenario will be valid when database is FHE encrypted.

Straightforward SQL injection may not be so severe when data is FHE encrypted and uploaded to the server. In case of any hacker executing encrypted query on the database, the returned result is also FHE encrypted and that will not reveal any information from database to the hacker. But, FHE data are not always being used by a single client, instead those can accessed by organizations with multiple access points. In this case, specific access control is required on different database columns to provide partial access to different users.

5.6 Conclusion

In spite of proposing different FHE-encrypted basic SQL constructs, it is expected that the response times of encrypted queries will be a bottleneck. Hence, improvement of underlying FHE library is a major requirement. Further, for application specific database design application-dependent implementations, optimizations and customized design approached can be taken to improve performance. Though emerging FHE technology is a new hope for future of cloud computing and other IT outsourcing models, a full-fledged FHE solution for encrypted databse is a long term goal for the industry. Here we primarily explain the design of general SQL queries with underlying FHE primitives. This primitives can further be improved with the use of future FHE libraries supported by hardware accelerators like GPU to obtain performances suitable for practical usage.

References

Ahn G-J, Sandhu R (2000) Role-based authorization constraints specification. ACM Trans Inf Syst Secur 3:207–226

Balduzzi M, Zaddach J, Balzarotti D, Kirda E, Loureiro S (2012) A security analysis of Amazon's elastic compute cloud service. In: SAC

Bernardo D, Assumpçao G (2009) Advanced SQL injection to operating system full control. In: Black hat Europe

Boneh D, Di Crescenzo G, Ostrovsky R, Persiano G (2004) Public key encryption with keyword search. Advances in cryptology – EUROCRYPT 2004, vol 3027. Lecture notes in computer science. Springer, Berlin, pp 506–522

Bradford Contel (2018) 7 most infamous cloud security breaches. StorageCraft Technology Corporation from https://blog.storagecraft.com/7-infamous-cloud-security-breaches/

Chatterjee A, SenGupta I (2018) Translating algorithms to handle fully homomorphic encrypted data on the cloud. IEEE Trans Cloud Comput 6(1):287–300

Chatterjee A, Manish Kaushal, SenGupta I (2013) Accelerating sorting of fully homomorphic encrypted data. INDOCRYPT 2013. Springer, Berlin, pp 262–273

Egorov M, Wilkison M (2016) ZeroDB white paper. CoRR abs arXiv:1602.07168

Felipe MR, Aung KMM, Ye X, Yonggang W (2015) StealthyCRM: a secure cloud crm system application that supports fully homomorphic database encryption. In: International conference on cloud computing research and innovation (ICCCRI)

FHE library HElib (2018). https://github.com/shaih/HElib

Grubbs P, McPherson R, Naveed M, Ristenpart T, Shmatikov V (2016) Breaking web applications built on top of encrypted data. In: CCS

Grubbs P, Ristenpart T, Shmatikov V (2017) Why your encrypted database is not secure. In: HotOS. pp 162–168

Guan Q, Zhang Z, Fu S (2011) Proactive failure management by integrated unsupervised and semi-supervised learning for dependable cloud systems. In: Proceeding 2011 6th international conference on availability reliability and security, ARES '11. Washington, DC, USA, pp 83–90

Huang C-T, Huan L, Qin Z, Yuan H, Zhou L, Varadharajan V, Jay Kuo C-C (2014) Survey on securing data storage in the cloud. ATSIP. https://doi.org/10.1017/ATSIP.2014.6

Juels A, Kaliski BS Jr (2007) Pors: proofs of retrievability for large files. In: Proceeding of 14th ACM conference on computer and communications security, CCS '07. pp 584–597

Kumar RS, Saxena A (2011) Data integrity proofs in cloud storage. In: 2011 3rd International conference on communication systems and networks (COMSNETS), pp 1–4

Lewko AB, Okamoto T, Sahai A, Takashima K, Waters B (2010) Fully secure functional encryption: attribute-based encryption and (hierarchical) inner product encryption. Advances in cryptology EUROCRYPT 2010, vol 6110. Lecture notes in computer science. Springer, Berlin, pp 62–91

Lillibridge M, Elnikety S, Birrell A, Burrows M, Isard M (2003) A cooperative Internet backup scheme. In: Proceeding of USENIX annual technical conference, ATEC '03. USENIX Association, Berkeley, CA, USA, pp 285–298

Lillibridge M, Elnikety S, Birrell A, Burrows M, Isard M, Ateniese G et al (2007) Provable data possession at untrusted stores. In: Proceeding of 14th ACM conference on computer and communications security, CCS '07. New York, NY, USA, pp 598–609

Miguel RF, Dash A, Aung KMM (2016) A study of secure dbaas with encrypted data transactions. In: Proceedings of the 2nd international conference on communication and information processing, ICCIP '16. pp 43–47

Popa RA, Redfield CMS, Zeldovich N, Balakrishnan H (2011) CryptDB: protecting confidentiality with encrypted query processing. In: Proceedings of the 23rd ACM symposium on operating systems principles (SOSP). Cascais, Portugal

Ristenpart T, Yilek S (2010) When good randomness goes bad: virtual machine reset vulnerabilities and hedging deployed cryptography. In: NDSS

Sahai A, Waters B (2005) Fuzzy identity-based encryption. In: Advances in Cryptology – EUROCRYPT 2005. 24th Annual international conference on the theory and applications of cryptographic techniques, Aarhus, Denmark, May 22–26. Lecture notes in computer science, vol 3494. Springer, Berlin, pp 457–473

Sandhu RS, Coyne EJ, Feinstein HL, Youman CE (1996) Role-based access control models. Computer 29(2):38–47

Storer MW, Greenan K, Long DDE, Miller EL (2008) Secure data deduplication. In Proceeding of 4th ACM international workshop on storage security and survivability, storageSS '08. New York, NY, USA, pp 1–10

Transaction processing performance council, TPC Benchmark H. http://www.tpc.org/tpc_documents_current_versions/current_specifications.asp. Accessed 22 Jul 2018

Verizon data breach incident report (2016). https://regmedia.co.uk/2016/05/12/dbir2016.pdf

Chapter 6
FURISC: FHE Encrypted URISC Design

As stated by Gosser, " *Securing a computer system has traditionally been a battle of wits: the penetrator tries to find the holes, and the designer tries to close them*". Hence, for any secure program execution, the instruction flow should also be encrypted. However, finding suitable solution to determine the termination point of any encrypted program is still an open challenge. The reason is encrypted termination requires handling of encrypted conditions with encrypted addresses, which is infeasible by existing unencrypted processors.

Thus, for outsourcing computations and achieving privacy, designs of processors which operate on encrypted data as well as address are extremely important. The conditions for designing encrypted processors are as follows:

- The modified arithmetic implemented in the processor must be a 'homomorphic image' of ordinary computer arithmetic;
- encrypted programs must never combine program addresses (the addresses of machine code instructions) with other data values;
- programs must be compiled either to save data addresses for reuse, or recalculate them exactly the same way the next time.

6.1 Existing Encrypted Processors

Secure data processing requires encrypted computations, which demands not only encrypted data but also encrypted address handling. This is the main motivation of designing encrypted processor. Table 6.1 shows existing encrypted processors in the literature. Existing literature shows that due to the performance bottleneck of FHE, in TsoutsosManiatakos (2013a, 2014) partial homomorphic encryption scheme has

© Springer Nature Singapore Pte Ltd. 2019
A. Chatterjee and K. M. M. Aung, *Fully Homomorphic Encryption
in Real World Applications*, Computer Architecture and Design Methodologies,
https://doi.org/10.1007/978-981-13-6393-1_6

Table 6.1 Existing encrypted processors

Architecture	Specification	Limitation
HEROIC TsoutsosManiatakos (2014)	Single instruction Partial homomorphic	Deterministic (Use of look up table) not CPA resistant
FURISC	Single instruction Fully homomorphic CPA resistant	Performance slow due to recryption
CryptoBlaze Irena et al. (2017)	Single instruction Partial homomorphic Performance faster	Performance highly dependent on decryption Server-side decryption
Pseudo-homomorphic Breuer et al. (2014) Processor	Pseudo (partial) homomorphic Super-scalar pipeline design	Encrypted operation in specific mode

been considered as a better choice while designing encrypted processor. In subsequent sections, brief details will be mentioned about existing partial homomorphic encryption (PHE) based crypto-processors. Subsequently, we shall discuss some limitations and highlight the requirement of designing FHE based processor.

6.1.1   Heroic: Partial Homomorphic Encrypted Processor

One of the first encrypted processor in literature is HEROIC (Homomorphic Encrypted One Instruction Computer), which utilizes single instruction architecture for processing encrypted data and Paillier homomorphic encryption is used as the underlying scheme to encrypt both instructions and data as mentioned in the work TsoutsosManiatakos (2013a, 2014). The HEROIC architecture builds upon the use of the Subtract and branch if negative (SBN) instruction, which requires three arguments (namely A, B and C), and its instruction is defined as:

```
Mem[B] = Mem[A] - Mem[B];
if (Mem[B] < 0)
    go to C
  else  go to  next instruction.
```

CryptoBlaze: Another PHE Based Processor

In recent work by Irena et al. (2017), CryptoBlaze is proposed as a partial homomorphic processor with eight specialized instructions and hardware to support computation of encrypted data. Main differences of CryptoBlaze with previously proposed HEROIC are:

- CryptoBlaze supports multiple instructions.
- It is based on non-deterministic Pallier encryption, hence in contrast to HEROIC few basic crypto attack (like chosen plaintext attack as discussed in Chap. 3) preventive.

CryptoBlaze is an extention of MicroBlaze instruction set architecture (ISA). MicroBlaze ISA is based on RISC (Reduced Set Instruction Computing) property. However, this architecture supports program memory which need to be unencrypted. This helps to support multiple instructions decoding, however imposes one challenge that program addresses and encrypted data should never be mixed. Hence, such type of processor is limited in execution power and can not support few instructions where program counter jump address need to be decided based on results of some encrypted operations (for example: go to).

Pseudo Homomorphic Encrypted Processor

Authors in Breuer et al. (2014) claimed to attain a processor where data is in encrypted form, however the processor is capable of working at near normal speed. Architecture of this design mainly works in two modes:

- **User mode**: In this mode, the processor runs on encrypted data with 32-bit instruction subset. The instructions are capable of accessing 32 general purpose registers (GPRs), and a very few permitted special purpose registers (SPRs).
  Supervisor mode: It supports more general instructions and pipeline, but supervisor mode processing is in unencrypted form.

In spite of these recent developments, these partial homomorphic or pseudo homomorphic processors have their own limitations. PHE based processors are either prone to some basic crypto attacks or supports some limited instructions. For pseudo homomorphic processor, change of mode limits the processing power. Another approach of designing encrypted processor considering FHE as an underlying scheme has been explained in Brenner et al. (2012a), which will be detailed in the subsequent sections.

FHE Based Encrypted Processor

Brenner et.al in (2012a) explained a framework to construct an encrypted multi-opcode using FHE as an underlying scheme. Use of FHE supports both encrypted addition and multiplication and hence encrypted decision making by FHE multiplexer. Thus, use of encrypted FHE multiplexer makes random program counter update and randomized memory access feasible. In this chapter, a modified FHE based processor will be discussed highlighting the major drawbacks of these existing encrypted processors.

Few implementations on homomorphic cryptosystems present in existing literature. Most of the implementations are either based on small parameters or make use of off-chip memory. Cao et al. in (2013) has presented integer-based FHE scheme implementation on a Virtex-7 FPGA (XC7VX980T). However, authors did not specifically measure the overhead for accessing off-chip memory and the design does not fit on current FPGA platforms. A 768K-bit multiplication based architecture is proposed in Wang and Huang (2013) based on Gentry-Halevi scheme GentryHalevi (2011a). A few other FHE based schemes are proposed in Wang et al. (2012, 2015) which works either in GPU platform or with multiple FPGA platforms. In Sinha Roy et al. (2015), authors have proposed an implementation of YASHE FHE Scheme Bos et al. (2013) with $n = 2^{15}$ parameter size, but authors do not consider the costs of moving data between external memory and the FPGA but assumed unlimited memory bandwidth. Hence, the design assumption is quite impractical. Contributions mentioned in Pöppelmann et al. (2015) consider the memory transfer time and optimize the algorithm in this regard. Thus Lattice based cryptosystems, which form the core units of FHE, face the challenge of large parameter sizes when implemented in hardware. In Doröz et al. (2013), FFT based million bit multiplication module has been proposed and full realization of FHE processor is proposed in Doröz et al. (2015) in 90 nm technology. However, realization of such processor requires storage of multiple large ciphertexts of million bits, which is still not feasible with existing memory sizes specifically for FPGA devices. Our final discussion will explore how basic building blocks for FHE processors can be implemented realized and suitably compressed, providing a feasible strategy to optimize the design for existing FPGAs.

In this context, our discussion justifies the usage of ultimate reduced instruction set (URISC) architecture to support operations on data and address encrypted by fully homomorphic encryption (FHE). The usage of URISC largely simplifies the opcode decoding compared to a conventional RISC based encoding on FHE based processors. Furthermore, the usage of FHE helps to provide resistance against Chosen Plaintext Attacks (CPA), which can be used to attack URISC based processors which operate on data and address encrypted with partial homomorphic encryption (PHE). The proposed architecture, named as FURISC, also reduces the amount of communication needed between a server on which the FHE operations are executed and the client which outsources the operations, thus providing a more effective solution to the termination problem while working with encrypted algorithms. A detailed architecture has been described in this chapter, comparing alternatives of using subtract branch if negative (SBN) and MOVE as FURISC primitives. Finally, a simulator of the SBN based FURISC architecture has been discussed to show that various algorithms can indeed be executed. Comparisons have been provided with a PHE based URISC to measure the overhead, while providing the much needed CPA security (as detailed in Chap. 2) due to the use of FHE.

6.2   Implementing Fully Homomorphic Encrypted Processor Using a Ultimate RISC Instruction

Ultimate RISC (URISC) is the minimalistic perspective to computer architecture design, where a single instruction is used to perform all computations. In this section, we first outline the rationale of using URISC for realizing FHE algorithms.

6.2.1 Justification of Encrypted Processor Along with Encrypted Data

Fully Homomorphic Encryption (FHE) provides an avenue for performing arbitrary computations on encrypted data. However, capability to operate on encrypted data alone is not sufficient for secured computation TsoutsosManiatakos (2014). In order to ensure that the control flow of the program is secured it is necessary that the address space is also encrypted. Consider, the program snippet, if(a[i] > a[j]) i = i + 1; It can be observed that if the data is encrypted, the outcome of the comparison is also encrypted which leads to the fact that, to update the index of the array the index also needs to be encrypted. Thus a processor architecture is necessary wherein the data and the memory content both are encrypted.

Requirement of encrypted processor has been explained in the notable work TsoutsosManiatakos (2014) from the security point of view. Use of cryptographic algorithms is one of the main solutions to maintain confidentiality in cloud. In this context, example of secured cloud microprocessor Fletcher et al. (2012) is mentioned, where the threat-model assumes that the pipeline is trustworthy. However, this security assumption becomes vulnerable if extraction of secured information is possible by eavesdropping the pipeline or the memory of the processor. A similar attack scheme in Becker et al. (2013) shows how to extract sensitive information from the chip with the help of sub-transistor level Trojan. To handle such threat models, processors should have the capability of executing encrypted instructions.

In the next section, we study the motivation of using a single RISC instruction, URISC, to build such an encrypted processor. We also address several related issues and discuss the motivation of choosing a FHE based URISC, which we call as FURISC.

6.2.2 Why URISC Architecture in Connection to FHE

For ensuring security and privacy in public clouds, it is important that the data and the functions computed there-on stay in an encrypted form. Thus the memory content, which stores both data and the instructions, have to be in an encrypted format. In Brenner et al. (2012b), authors have given an initial layout of how to handle such encrypted instructions using FHE encrypted multi-instruction processor. However, we highlight that the problems of using multi opcodes in FHE based processor is detailed as follows in case of performance based challenge:

6.2.3 Performance Based Challenge

For multi-opcode processor, the execution time of any particular instruction, the (*Instruction_Time*) is the summation of *Encrypted opcode search time* and actual execution time of the instruction. By *Encrypted opcode search time*, we mean time to search the opcode match between the opcodes present in the program counter and opcodes in the instruction set.

$$Instruction_{Time}(T) = Encrypted\ opcode\ search\ time + Execution\ Time \qquad (6.1)$$

Now, we outline how the search is actually performed using the homomorphic modules explained in Chap. 4:

- Let an encrypted data s is to be searched in a database D with n encrypted data items.
- Initially, FHE subtraction (using *FHE_Sub* module) operations are performed between s and each of the database items. Let the encrypted subtraction results be $sub_0, sub_1, \ldots, sub_n$.
- Each subtraction result is checked if it is equal to 0 (using *FHE_Equal* module). Any subtraction result sub_i equals to 0 implies that s matches with the ith item of the database. If no subtraction result equals to 0, it indicates s is not present in the database.
- However, it requires decryption of n outputs of *FHE_Equal* module to take decision about the search result at this stage. To avoid such large number of decryptions, we perform post processing on the outputs of *FHE_Equal* module in encrypted domain to generate a single bit encrypted search result. If the search result is Enc(1), it indicates the search item matches with one or more items present in the database.
- In the post processing phase, each of the outputs of the *FHE_Equal* module is inverted using *FHE_Inv* module and bit-wise FHE multiplication is performed among the inverted bits. If any of the outputs of *FHE_Equal* module equals to 1 (indicates match is found), then it inverts to bit 0. All other outputs of equality check becomes 0 and inverts to 1. Thus, FHE multiplication result bit becomes 0 (or 1), depending on any match is found or not.
- The inversion of this multiplication result is stored directly in the server as the encrypted search result. If the final multiplication result is 0, it indicates the search is unsuccessful and the intended data is not present in the database.

The required submodules for FHE search were implemented as mentioned in Sect. 4.4. The above mentioned encrypted search algorithm has been evaluated for correctness on a Linux Ubuntu 64-bit machine with 1.6 GHZ clock and 8 GB RAM. Figure 6.1 shows the timing requirement for searching which shows for a significantly large n, the search time is quite high. Further, Table 6.2 shows the number

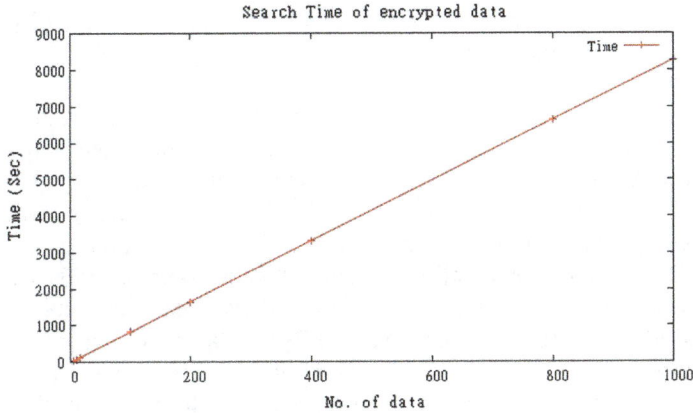

Fig. 6.1 Timing requirement for FHE Search

Table 6.2 Number of instructions required for different operations in URISC

	Operations	Required SBN instructions
1.	Addition	3
2.	Subtraction	1
3.	Jump	1
4.	Move	4
4	Branch	2
5.	Not	10

of instructions (k) required to implement one operation in OISC is notably less than *Instruction_Time*(T) for a decent number of instructions n. This shows the idea of multi-instruction FHE based processor in Brenner et al. (2012b) has large search time overhead and this is our main motivation to choose OISC architecture while implementing FHE based processor.

OISC provides a unique opportunity in this context. An OISC is an abstract machine, which uses only a single instruction and other necessary instructions are composed from the single instruction set Gilreath and Laplante (2012). Thus, the OISC is *Turing Complete*, and one can perform all computations using a single instruction. This resolves the confusion regarding varying opcode in case of a standard RISC or CISC processor, which has multiple instructions in their ISA.

6.2.3.1   Pitfalls of Using Partial Homomorphic Schemes: Why Fully Homomorphic Encryption?

For designing encrypted processors, partial homomorphic schemes are the first choice since FHE schemes suffer from performance issues. Following the contributions in TsoutsosManiatakos (2013b, 2014), authors have explored the design of encrypted one instruction set processor based on the Paillier based encryption, which is an additive homomorphic encryption scheme. The underlying instruction SBN (mentioned in Sect. 6.1.1) is a single instruction whose arithmetic computation is a subtraction on two operand values. In the same instruction, depending on whether the result is positive or negative, the Program Counter (PC) gets updated to the next address or an instruction mentioned as another operand of the SBN instruction. Since, Paillier encryption supports subtraction on encrypted data this is a promising choice to develop a processor for performing computations on encrypted data (decompose the program using encrypted SBN instructions, and subsequently execute them using Paillier encryption algorithm as an underlying scheme).

Unfortunately, the design suffers from a serious deficiency. The PC needs to be updated based on an encrypted condition after the subtraction. Thus while the subtraction is supported by the underlying partial homomorphic encryption scheme, the PC update needs an encrypted decision making module which can be realized by an encrypted multiplexer. To explain, consider a decision block, where depending on an encrypted condition c' (complement denoted by $\overline{c}'$), output y' may be a' or b' (all the variables are encrypted). The decision block can thus be realized by a multiplexer $y' = a'(c') + b'(\overline{c}')$, where the computations of the right hand side are homomorphically performed. This design of a multiplexer on encrypted data and encrypted control requires capability to perform both encrypted addition and multiplication, which is not supported by any partial homomorphic scheme.

To make these decisions, the design proposed in TsoutsosManiatakos (2014) uses lookup memory table for storing sign for encryptions of numbers. Moreover, it is assumed that the encryption is deterministic and the public key of the encryption is unknown to the adversary. In several real life scenarios such a restriction may not be feasible and the deterministic encryption makes the processor computations vulnerable to chosen plaintext attack (CPA) Katz and Lindell (2007).

This motivates us to look into replacing this underlying partial homomorphic encryption scheme with FHE, which in turn is capable of designing an encrypted decision module, namely the multiplexer. This provides the flexibility of making branch decisions and PC update even when the encryption is randomized, without the use of any static encryption table or deterministic encryption.

6.3   Design Basics of FURISC

There are different types of single instruction based models like subtract and branch if less than or equal to zero, subtract and branch if negative (SBN), reverse subtract and skip if borrow, Move. We consider the FURISC architecture with SBN as the underlying scheme. More formally, the instruction in 3-tuple form is represented as:

```
SBN A, B, C :
   Mem[B] = Mem[A] - Mem[B];
        if (Mem[B]< 0)
              PC <- C
        else  PC <-  next instruction
```

In this paradigm, `Mem[B]` is computed homomorphically by `Mem[A]` – `Mem[B]`. Depending on whether the value of `Mem[B]` is negative, homomorphically the PC is updated. The PC eventually gets the value of End of Program (EOP), when the program terminates. In the subsequent sections, we present a detailed description of the FURISC architecture.

6.3.1 Design of FURISC

In this section, we discuss the design basics of FURISC processor based on two primitive URISC instructions: SBN and MOVE. Direct implementation of FURISC processor based on SBN instruction incurs overhead in terms of handling different operations. Table 6.3 shows few operations which require large number of single instructions. Specially, we focus on *Move* and *Not* operations, which require 4 and 10 single instructions respectively. Mapping of any operation to large number of single instructions lead to large memory handling, which is costly operation. That inturn diminishes the main objective of using URISC instruction for performance enhancement. Hence, we modify the basic SBN instruction by adding the parameter *resultant'*. Using this format, *Move* instruction is realized in the following way:

$$resultant' = Mem'[A'] - enc(0); \qquad (6.2)$$

Content of $Mem'[B']$ is loaded with encrypted 0, hence the content of $Mem'[A']$ is moved to *resultant'* by single subtraction operation.

Similarly with the same format, *NOT* operation is realized in the following way:

$$resultant' = enc(0) - Mem'[B']; \qquad (6.3)$$

Content of $Mem'[A']$ is loaded with encrypted 0, hence *NOT* of $Mem'[B']$ is stored to *resultant'* by single subtraction operation.

Here we consider 4-tuple format of SBN instruction and explain how to design a SBN based FURISC.

Let A', B' and C' be FHE encrypted memory addresses and Mem'[A'] and Mem'[B'] be the encrypted contents of the respective addresses. With these parameters, fully homomorphic SBN instruction can be represented as:

```
SBN A', B', resultant', C' :
  resultant' = Mem'[A'] - Mem'[B'];
    if (resultant' < enc (0))
          PC' <- C'
    else  PC' <- next instruction
```

Implementation of this FHE based SBN instruction requires the following steps:

- *Encryption Phase*: Memory addresses A, B, C should be encrypted by FHE to A', B', C' and contents of the addresses are stored in encrypted format.
- *Memory read-write*: Contents of memory address A' and memory address B' need to be fetched. *Encrypted memory module* handles memory read and write operation which will be explained in a subsequent section.
- *FHE_Subtraction*: Subtraction of Mem'[A'] and Mem'[B'] is performed by FHE_Sub module of FURISC processor and stored in register resultant'.
- *Branching or program counter (PC) Update*: If (resultant' < enc(0)), the execution control proceeds to C' or to next instruction pointed by encrypted PC i.e (PC' + 1'), where 1' is the encryption of 1. This branch update is handled by *FHE_Branch* module, which is again part of Encrypted ALU of FURISC.
- Value of the register resultant' is finally updated to certain memory or register address location according to URISC instructions.

6.3.2   *Encrypted Memory Module*

Encrypted memory in FURISC design requires manipulation of encrypted data as well as encrypted addressing. The main design challenge of designing such memory is that the underlying encryption algorithm is randomized. Hence, initially encrypted data may be stored in a certain encrypted address. During memory-fetch encryption of same address gives a different result (bitwise values are different). That makes the content fetching more difficult from a particular address of memory. Hence, an encrypted decision making module is required for encrypted memory read-write as proposed in Brenner et al. (2012a).

However, performing linear search on large encrypted memory requires large timing requirement. Hence, we perform memory search in parallel fashion for each of the parameters. Revisiting the SBN instruction: *SBN A', B', resultant', C'*, PC update requires memory comparisons for each of the parameters *A', B', C'* as well as *resultant'* and this memory search is performed in parallel to improve the performance.

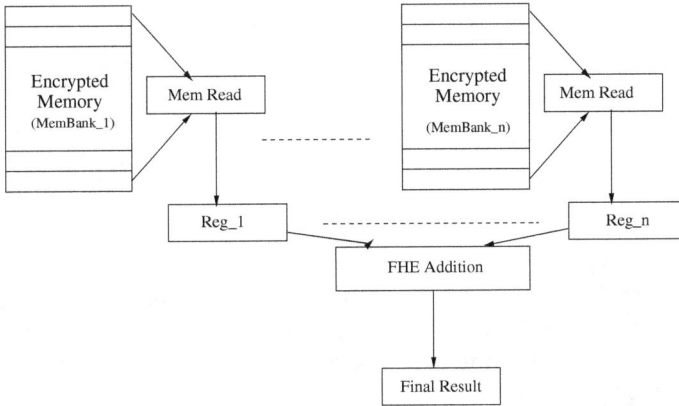

Fig. 6.2 Interleaved memory architecture

In our implementation, we have further used the interleaved memory format. The overall memory is divided in different memory banks and memory search is performed in each of the memory banks as shown Fig. 6.2. For 1 GB memory representation, each of the memory banks are divided into 100 MB to perform faster search. Let there be n memory banks and search result of each memory bank is stored in Reg_i, where $0 \leq i \leq n$. By the search result, we mean if any memory read operation is performed and address match is found in ith memory bank, the memory content is stored in Reg_i, otherwise Reg_i holds encrypted 0. Finally, all the Reg_i contents are added to get the memory read value. This modification of two stage memory design improves the performance of overall read operation. Next in Sect. 6.3.3, we shall discuss the design of encrypted ALU module for FURISC.

6.3.3 Encrypted ALU Module

The main arithmetic operations of this ALU for FURISC processor is FHE subtraction and PC update as shown in Fig. 6.3. The ALU module mainly consists of a fully homomorphic subtraction module (**FHE_Sub**) and FHE branch module (**FHE_branch**).

FHE_Sub module: FHE Subtraction is implemented by adding one number with the 2's complement of another. The subtraction module is designed by performing homomorphic addition of one ciphertext with 2's complement of another ciphertext.

FHE_branch module: According to the principle of SBN instruction, branching operation decides whether the program control will next proceed to address C′ or to the next address of program counter (PC′ + 1′). Since all the operations will take place in encrypted domain in FURISC, the next proceeding address should also be encrypted. For this reason, FHE_MUX is used with two inputs, C′ and the

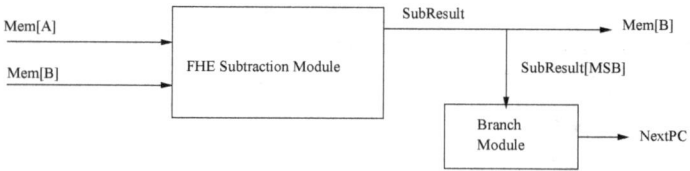

Fig. 6.3 Encrypted ALU module for FURISC

incremented `PC' + 1'`. The branching depends on the decision if the subtraction result of `Mem'[A']` and `Mem'[B']` is negative. Hence, the most significant bit (MSB) of the subtraction result is treated as the selection line ($MSB = 1'$ indicates the value as negative).

6.3.4 Overall Architecture

Figure 6.4 shows the overall architecture with encrypted memory module and encrypted ALU. SBN functionality is realized with the following steps with this architecture:

- Register `A'`, `B'` and `C'` hold the address values as mentioned in the SBN instruction parameter.
- Initially, address of `A'` is taken into `PC'` and the memory content is fetched from the `Encrypted Memory` by memory read operation. `Memory Read/Write Module` works as mentioned in Sect. 6.3.2. The fetched value is stored in register `Mem'[A']`.
- Similarly, contents of memory address `B'` is stored in `Mem'[B']`. Selection of `A'` or `B'` is controlled by `sel`, the selection line of associated FHE_MUX.
- Subtraction operation is performed using the `FHE ALU` module and the result is stored in `Resultant` register.
- Further, MSB of `Resultant` register value is fed as selection to a FHE_MUX for PC update and the next PC address is determined from the two inputs (`PC' + 1'`) and `C'` of the multiplexer depending the selection value. It may be noted that (`PC' + 1'`) can be obtained by homomorphically adding the cipher corresponding to 1 with that corresponding to PC.
- Depending on the third parameter of the SBN instruction, value stored in the `Resultant` register is updated in the respective memory or register location.

Table 6.3 shows the number of required CPU cycles to implement different encrypted functions on encrypted SBN and Move based FURISC. The results are obtained designing C-based simulators of SBN based FURISC architectures. The simulators are designed and evaluated for correctness on Intel(R) Xeon(R) CPU E5-2697 v2 2.70 GHz with $2 * 12$ cores and 256 GB RAM. Among the implemented functions, Fibonacci requires single loop handling, binary search and sort algorithms

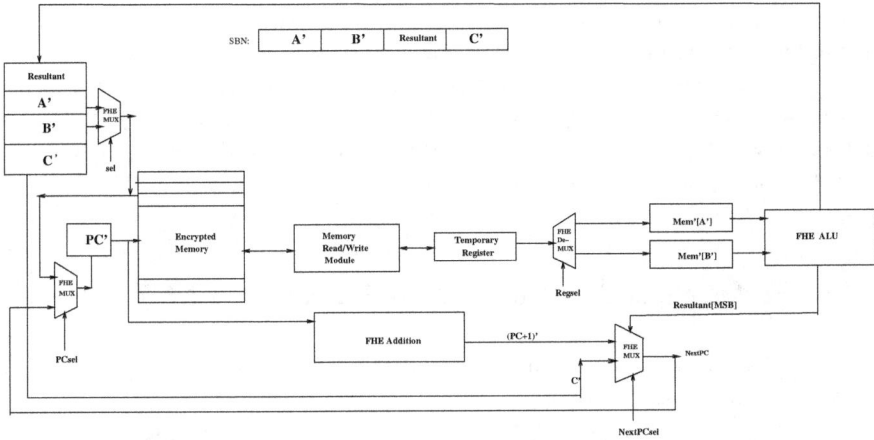

Fig. 6.4 Overall FURISC architecture

Table 6.3 Timing requirement of encrypted operations on FURISC

Operations	CPU cycles
SBN processor implementation	
	Scarab timing
Fibonacci	$3918 * 10^8$
Binary search	$96 * 10^8$
Quick sort	$12012 * 10^8$

require multiple loop handling. Here, we show the required CPU cycles for computing Fibonacci value of 100, for performing binary search within 100 data and performing quick sort on a collection of 100 data.

So far we have discussed how to design SBN based FURISC. Another approach of FURISC design is based on MOVE instruction, which basically works on copy operation. Actual computation of MOVE based architecture takes place by the underlying memory mapped hardware. In the next section, we outline a comparison between MOVE and SBN based FURISC and explore which design is actually advantageous in terms of performance.

6.4   Comparison with MOVE Based URISC

The format of the basic instruction for MOVE based FURISC is:

```
MOVE operandam' operandum'
```

The implication of this instruction is to copy the contents of the operandam' to operandum', where these are two encrypted addresses. The copy can be performed from any location to other (to any memory or register from any memory or register). Hence, the design of a MOVE based architecture only requires memory fetch-write operations and register fetch-write operations. Memory read-write operations are performed using the modules explained in Sect. 6.3.2.

6.4.1 Performance Evaluation: SBN Versus Move FURISC

In this section, we evaluate the FURISC architecture alternatives between SBN and MOVE in terms of performance. According to present literature, realizing of basic building blocks to implement FHE based processor in actual hardware is another major challenge due to large parameter sizes as shown in Doröz et al. (2015). Hence, in this work we concentrate on software simulation results. Here, SBN and MOVE based implementations are compared in terms of number of instructions and it is investigated which one is better in terms of performance.

Let a program P be implemented by n_1 SBN instructions. Let same program P can be implemented by n_2 MOVE instructions. Now, let a single SBN instruction be implemented by m_1 MOVE instructions. Hence, intuitively converting all SBN instructions of program P to MOVE instruction is equivalent to implementing P only with MOVE instructions. That implies the code length of the program P using only MOVE instructions is proportional to $n_1 m_1$ MOVE instructions. Similarly, let single MOVE instruction be implemented by m_2 SBN instructions, hence $\frac{n_1 . m_1}{n_2 . m_2} = \frac{n_2}{n_1}$ or $(\frac{n_1}{n_2})^2 = \frac{m_2}{m_1}$. Following code snippets show how single SBN and MOVE instructions can be mapped to their respective MOVE and SBN equivalents.

A single MOVE instruction MOVE operandam operandum can be realized by a single SBN instruction:

```
SBN operandam, #00, operandum, #00
```

However, a single SBN instruction: SBN operandam operandum resultant next-address can be realized by the following instructions:

```
INVERT operandum
ADD operandam operandum resultant
COMPARE resultant CONSTANT
BRANCH next-address
```

In these instruction sequences, operandum is inverted and (-operandum), added to operandam and the addition result is stored in resultant. The value of resultant is compared with CONSTANT (which is zero) to check if the value is negative and branch to next-address depending on the value of the resultant.

All the instructions like INVERT, ADD, COMPARE, BRANCH can be realized by multiple MOVE instructions Gilreath and Laplante (2012). Hence, number of MOVE instructions required to implement a single SBN instruction (m_1) is greater than number of SBN instructions equivalent to one MOVE operation (m_2) Gilreath and Laplante (2012). That again implies, $m_1 > m_2$ and hence $n_1 < n_2$. Hence, it indicates SBN based URISC architecture requires lesser number of instructions compared to MOVE instruction based URISC to implement any program. In practical scenario, large number of instructions indicate large number of PC update and memory, register handling. In both SBN and MOVE based FURISC architecture, PC update and memory and register read-write operations require large number of FHE operations, hence that incurs extra timing requirement in terms of CPU cycles. Due to this reason, MOVE based FURISC is not advantageous in terms of performance compared to SBN. Hence, in the subsequent sections, we shall only consider SBN based FURISC architecture. In the next section, we explain how encrypted programs can be realized using FURISC.

6.5 FURISC Applied to Realize Encrypted Programs

In this section, we explain with examples how FURISC architecture tackles encrypted loop execution. Initially, we start with an example of a simple loop:

```
while(x > y)
  {
      x--;
  }
```

Since, x and y are both encrypted, the termination condition of the loop while (x > y) is impossible to comprehend when the code is executed in any general purpose processor. In FURISC architecture, SBN instructions realize the loop in the following way:

```
while' : SBN $1000', $1001', temp', &wend'
         SBN $1000', Reg1, $1000', null
         // Reg1 stores the value of enc(1)
         SBN PC', &while', PC', &while'
wend'    SBN $1000', Reg0, $1000', null
         // Reg0 stores the value of enc(0)
```

Let encrypted x and y be stored in encrypted addresses $1000'$ and $1001'$. The starting encrypted address of the while loop execution is denoted by while' whereas wend' denotes the encrypted address where PC should jump once the while loop terminates. The advantage of designing this FURISC is that PC update can be controlled by encrypted subtraction operation since PC and all the address locations are encrypted. Thus, when encrypted x is less than encrypted y, subtraction result between contents of $1000' and $1001' becomes negative. That indicates the encrypted termination condition has been reached and PC now should jump to wend'. If this is the only loop present in program then the wend' is a no operation (NOP) and PC next jumps to *End of program* location. Otherwise, PC jumps to next instruction of the program. NOP is implemented by subtracting enc(0) from the value of $1000' location and storing the result back to the $1000' location.

In the next example, we shall show how multiple encrypted loop termination is handled using FURISC. We take the example of Quick sort which consists of multiple nested loops. Consider the following pseudocode of Quick sort. The variables are self explanatory corresponding to classical Quick sort.

```
1.  if(first<last){
2.              pivot=first;
3.              i=first;
4.              j=last;

5.          while(i<j){
6.              while(x[i]<=x[pivot]&&i<last)
7.                      i++;

8.                  while(x[j]>x[pivot])
9.                      j--;

10.                 if(i<j){
11.                     temp=x[i];
12.                     x[i]=x[j];
13.                     x[j]=temp;
14.                 }
15.             }

16.             temp=x[pivot];
17.             x[pivot]=x[j];
18.             x[j]=temp;

19.             quicksort(x,first,j-1);
20.             quicksort(x,j+1,last);
21.     }
22. }
```

Following is the representation of this code realized with SBN instruction architecture. The code has been accompanied with comments while relating it to the classical Quick sort. The comments also show the relationships of the code with the lines of the pseudocode.

```
/********* FURISC implementation
             of Quick sort   *********/
/***** Starting of if: lines 1-4 ******/
QS':     SBN last', first', temp', &EOP
         SBN first', Reg0, pivot', null
       // Reg0 stores the value of enc(0)
         SBN first', Reg0, i', null
         SBN last', Reg0, j', null

/********* while(i<j): line 5 *********/
while1': SBN j', i', temp', &wend1'

         SBN Reg0, pivot', jtemp', null
         SBN $2000', jtemp', temp1', null

/********* while loop : line 6-7 ******/
while2': SBN Reg0, i', itemp', null
         SBN $2000', itemp', temp', null
         SBN temp1', temp', accumulator',
                                    &while3'
         SBN last', i', product, &while3'
         SBN product, Reg0,temp',&while3'
         SBN Reg0, Reg1, temp', null
       // Reg1 stores the value of enc(1)
         SBN i', temp', i',null

/********* End of while of line 6  ******/
wend2'   SBN PC', &while2', PC', &while2'

         SBN Reg0, pivot', jtemp', null
         SBN $2000', jtemp', temp1', null

/******** while loop : line 8-9 *******/
while3': SBN Reg0, j', jtemp', null
         SBN $2000', jtemp', temp', null

         SBN temp1', temp', accumulator',
                                    &wend3'
         SBN j', Reg1, j'
         SBN PC', &while3', PC', &while3'

/******* End of while of line 8 *******/
wend3': SBN j', i', temp', &endif'
```

```
            SBN Reg0, i', itemp', null
            SBN $2000', itemp', temp', null
            SBN $2000', itemp', (mem_tempi)',
                                         null
            SBN Reg0, j', jtemp', null
            SBN $2000', jtemp', temp1', null
            SBN $2000', jtemp', (mem_tempj)',
                                         null

            SBN  temp', Reg0, (mem_tempj)',
                                         null
            SBN  temp1', Reg0, (mem_tempi)',
                                         null


endif : SBN PC', &while1', PC', &while1'



/****** End of while of line 5 *******/
wend1': SBN Reg0, pivot', itemp', null
            SBN $2000', itemp', temp', null
            SBN $2000', itemp', (mem_tempi)',
                                         null
            SBN Reg0, j', jtemp', null
            SBN $2000', jtemp', temp1', null
            SBN $2000', jtemp', mem_tempj',
                                         null

            SBN  temp', Reg0, mem_tempj',
                                         null
            SBN  temp1', Reg0, mem_tempi',
                                         null


            SBN j', Reg1, j', null
            SBN QS', Reg0, PC', PC'


            SBN Reg0, Reg1, temp', null
            SBN j', temp', j'
            SBN QS', Reg0, PC', PC'

/********* End of Program *********/
EOP: SBN $2000', Reg0, $2000', null
```

This code snippet shows how the nested loops can be handled using this architecture. Here, we consider an array x[ ] is resided at starting address $2000'. At while1', the condition (i' < j') (i' and j' are the encryptions of i and j) has been checked using (SBN i', j', temp', wend1'), where i' and j' are stored in intermediate registers. With the SBN functionality, j' is subtracted from i' and the loop condition is checked. When j' is less than i', subtraction result is negative and PC' proceeds to end of while (wend1'). For while2', loop condition is checked by subtracting i' from last' and if the subtraction result is negative the program flow is branched to while3'. Thus, multiple loops are handled without the requirement of any redundant operations. Once, the termination condition is reached PC should jump to the *End of program* location. A processor which operates on FHE data suffers from the challenge of identifying the termination points of its control logic. Typically, a program consists of multiple loops, where the loop termination conditions are based on checks which operate on encrypted information homomorphically. Because of this uncertainty, the FHE programs when transformed in a straight forward manner from their unencrypted versions, are unsure whether to iterate or to move towards the next statements. However, one may note that one way to ameliorate this issue is by observing that if the program is written in the circuit representation in URISC format, then final terminal point of the program can be determined by the worst case analysis of the algorithm, since FURISC works based on the circuit representation of the algorithm Goldwasser et al. (2013b).

6.6 Results

Results in proposed work Brenner et al. (2012a) mention only the ALU execution and single machine cycle execution timing. However, overall instruction execution timing considering opcode handling, memory read-write are not provided.

Table 6.4 shows a comparison of FURISC performance with PHE based HEROIC proposed in TsoutsosManiatakos (2013b, 2014). According to the experimental results, FHE based FURISC requires more clock cycles compared to HEROIC for implementing the same operation, but it is advantageous in terms of secu-

Table 6.4 Comparison of FURISC performance with HEROIC

Operations	HEROIC timing (Clock cycles)	FURISC timing (Clock cycles)
		Scarab timing
Factorial	$8.45 * 10^7$	$402.5 * 10^8$
Fibonacci	$2.74 * 10^8$	$396 * 10^8$
Bubble Sort	$1.54 * 10^8$	$3509 * 10^8$

Table 6.5 Encrypted
operations timing using
FURISC

Operations	Timing (min)
Addition	63.87
Subtraction	21.3
Jump	21.3
Move	21.3
Branch	42.8
NOT	21.3

rity improvement and providing CPA resistance to the encrypted processor. Unlike TsoutsosManiatakos (2014), access to public key need not be restricted and the encryption can be randomized for CPA resistance.

All our computations are implemented in Intel(R) Xeon(R) CPU E5-2697 v2 2.70 GHz with $2 * 12$ cores and 256 GB RAM. Proper parallelization techniques are used for memory module implementation in this work. Our comparison result shows a hope that in spite of the fact FHE based OISC is much slower compared to PHE based designs, a faster underlying FHE library (specially using parallelization techniques for the overall underlying library as well) in future can indeed make the FURISC design practical.

However, library libScarab (2011) depends on smaller security parameters and normal FHE schemes with higher security parameters suffer from performance issues and still impractical to use. Leveled FHE can evaluate circuits with bounded depth faster, where it is sufficient to evaluate the encrypted multiplexer as well as other basic building blocks of FURISC. Hence, performance of this FURISC architecture can be estimated in reconfigurable hardware platform considering YASHE as the underlying scheme (for the parameter set $n = 16384$, $log_2(q) = 512$ and capable of evaluating 9 levels of multiplications) in Virtex-7 XC7V1140T FPGA platform. As mentioned in the work by Pöppelmann et al. (2015) the FHE scheme requires 0.94 s for homomorphic addition and 48.56 s for homomorphic multiplication. Based on this result, Table 6.5 shows the timing estimates for different standard operations.

6.6.1   Drawback of FURISc

We have discussed an encrypted URISC architecture with FHE as underlying encryption scheme, which combines the flexibility of performing arbitrary operations on encrypted data due to the property of FHE with design simplicity of URISC architecture. Due to the use of FHE, randomization in memory handling and PC branching solves the CPA Vulnerability issues of previous PHE based design. Further, we also observe how this design is advantageous to handle encrypted loop termination problem. However, if we closely follow the structure of URISC instruction (

```
SBN A', B', resultant', C'
```

), one drawback of such URISC based design is high memory read-write overhead. To access the memory locations for the URISC parameters, it is required to traverse and compare throughout the whole memory module each time, which is very performance costly for FHE encrypted memories. Hence, it leads to a scope of future work to study how designing of application specific complex instruction set based FHE processor may confirm security and improve performance by reducing significant memory access and opcode search overhead.

Performance improvement of FURISC is the main bottleneck to realize this design in practical. Design of a dedicated FHE based processor in hardware may be beneficial in case of improving performance. Efficient implementation of homomorphic encrypted processor with a proper balance between security, performance, and key as well as ciphertext sizes is a challenging area of research Pöppelmann et al. (2015) Pöppelmann et al. (2015). First full realization of FHE in hardware was reported in Doröz et al. (2015) that is synthesized using 90 nm technology. In Roy et al. (2017), a scheme has been proposed to a scheme to perform homomorphic evaluations of arbitrary depth with the assistance of a special recryption box. Work in Cousins et al. (2017) contributes further in this direction by developing FPGA based hardware primitives to accelerate the computation on encrypted data using a specific FHE cryptosystem based on NTRU-like lattice techniques with efficient key switching and modulus reduction operations to reduce the frequency of bootstrapping.

In the next section, we shall explore the challenges of actually realizing such FHE based processor in reconfigurable hardware.

6.7 FHE Processor Implementation Challenges

RSA Rivest et al. (1978b) and ECC cryptosystems Koblitz et al. (2000) are the main pillars of existing classical cryptosystem based on factorization and the elliptic curve discrete logarithm problem (ECDLP). However, Shor's algorithm Shor (1994) has shown a threat that underlying mathematical assumptions of these classical cryptosystems can be solved by quantum computer. Hence, present research focus on actual hardware implementations of post-quantum secure solutions like homomorphic encryption.

However, recently proposed encryption schemes like much-discussed FHE and other digital signature schemes based on lattice Lyubashevsky (2008); Lyubashevsky et al. (2010); Lindner and Peikert (2011); Lyubashevsky (2012) demand large parameter sizes. For efficient implementations to be used in practical scenarios, use of ideal lattice helps to reduce the parameter size. Ideal lattice is the backbone of somewhat homomorphic encryption (SHE) and fully homomorphic encryption (FHE) schemes.

Implementing lattice based homomorphic cryptosystems on reconfigurable platform is a challenging issue due to large key and ciphertext sizes. Existing works either target to use large amount of external memory or work on smaller parameters.

In this chapter, we explore the use of run length encoding to represent the large ciphertexts in a compressed way and modify addition and multiplication algorithms to be executed on encoded data in a memory efficient manner.

6.8   Utilizing Compression Technique for FHE Architecture

Reconfigurable computing demands an efficient configuration with minimal amount of data transfer to be effective. However, realization of homomorphic processor requires storage of multiple large ciphertexts of million bits, which is still not feasible with existing memory sizes of FPGA devices. Sending this large amount of information to the FPGA from external memory can be quite time and power consuming. A logical solution would be to compress the data stream, which will in turn reduce the amount of external storage needed to hold the configuration, reduce the amount of time needed to send the configuration information to the device, and reduce the amount of communication through the power-hungry off-chip I/O of the FPGA. Hence, a proper compression technique to store such large parameters in comparatively smaller memory can be helpful to make such implementations actually feasible.

The required ciphertext compression for FPGA must be lossless. As a compression technique, we mention the use run length proposed by Nelson (1991) and subsequently work on how to perform addition and multiplication on huge ciphertexts in encoded form.

6.8.1   Run Length Encoding

Run length encoding (RLE) is a lossless compression technique, where a sequence is stored in a special form. Sequences in which the same data value occurs in many consecutive data elements, those are represented with single data element value and the count of repetitions of data value Hauck and Wilson (1999). The advantage of this scheme is to represent n consecutive same data bits, RLE encoded form requires maximum ($\log n$ bit (to store the count) $+1$ (to store the bit value)).

6.8.2   Proposed Encoding Module

In this section, we discuss the design of encoding module to store large ciphertexts in a memory efficient way. Due to randomization of encryption scheme, generated ciphertexts typically consist of subsequences in which the same bit value occurs consecutively. This repetition gives an opportunity to apply a compression technique and store only the bit value and count of the number of repetitions in each of the subsequences. This run length encoding can be used to store each ciphertext. To store every single subsequence with repetition of same bit value, single dedicated RAM block with $(n + 1)$ bit is used. n-bit is used to store the subsequence length and extra 1-bit is to indicate the bit value being repeated.

Figure 6.5 explains the encoding technique with an example. In the example, 3-bit *cnt* of RAM stores the length of sequences (count value shown in diagram as *cnt*) of same bit values (considering maximum possible consecutive count can be stored as 7 in a single block of RAM) and 1-bit *btvalue* stores the value of the bit (1 or 0) in the RAM. In the rest of the chapter, we mention the sequence length with consecutive same bit values as *sublength* and the ith sequence as *block$_i$*.

Figure 6.6 shows the hardware module to encode a single ciphertext (or any large parameter in homomorphic processor like key). Initially the ciphertext can be stored in the ROM or it is being generated from some previous intermediate addition or multiplication operations. The ciphertext bits are fetched gradually and it is fed to the *Count_bit module*. Further, *Count_bit module* counts the number of consecutive same bits, and sets the *bitout* value as the repetitive bit value and stores the count in *bitcount*. Once the bit value changes (sequence of 1 finishes and bit value 0 occurs or sequence of 0 finishes and bit value 1 occurs), it sets the *RAM wen* high to store the *bitout* and *bitcount* value in a RAM block.

However, field addition and multiplication need to be performed while implementing the actual functionality of cryptoprocessors. Direct addition or multiplication approaches are not applicable while working with large ciphertexts, which is stored in the encoded representation. Hence, we need to modify our algorithms in such a way that most of the ciphertext bits remain in the encoded format before

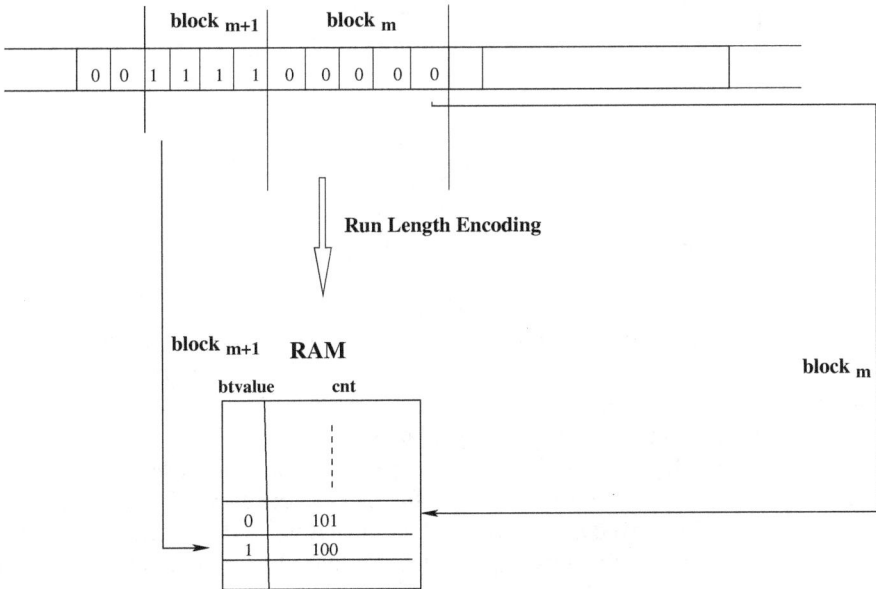

Fig. 6.5 Encoding example

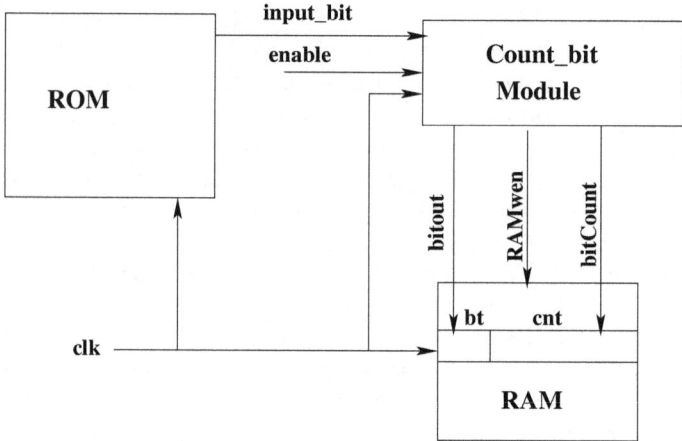

Fig. 6.6 Encoding architecture

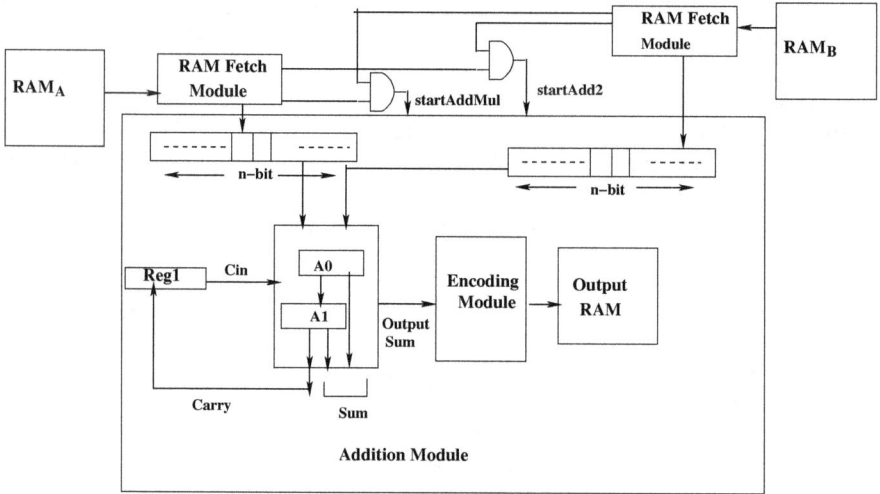

Fig. 6.7 Encoded addition module

addition (or multiplication) can be done on smaller bit sizes at a time. Let two large ciphertexts A and B be stored in RAM_A and RAM_B (both are dual output port RAMs). Figure 6.7 shows how to perform addition between A and B by fetching block of bits from both the RAMs and decoding them gradually.

However, for RAM_A and RAM_B, each of the blocks will not have the same *cnt* value, since subsequences of each of the ciphertexts are different. Hence, addition is performed by fetching fixed *n*-bits from each of the RAMs. A dedicated *RAM_Fetch module* fetches *n*-bits of the operands and transfer the same to addition module (or multiplication module,). Addition is performed on the outputs of the *RAM_Fetch*

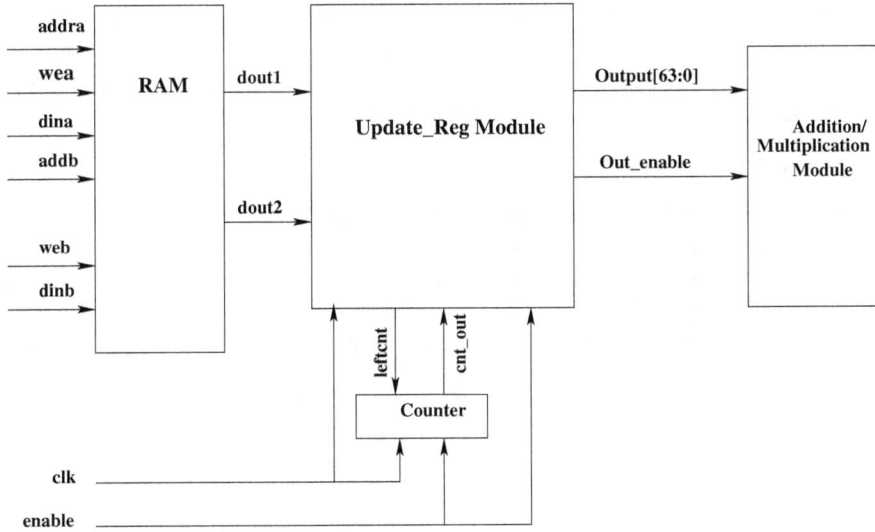

Fig. 6.8 RAM fetch module

module from RAM_A and RAM_B. Figure 6.7 shows the addition hardware, which consists of:

- **RAM_Fetch module**: This module fetches data from consecutive RAM blocks and it is fed to the *Update_Reg Module*. The *Update_Reg Module* consists of a counter input and counts if the number of the fetched data-bits has been reached to fixed n. During the fetch, let the total count of number of fetched bits at any point be k. If the count is such that $k < n$, next RAM block is fetched. Let l indicate the count of the fetched bits from the next RAM block and if $k + l > n$, then the *Update_Reg Module* stores the rest $k + l - n$ bits to be processed during the fetch of next n-bit. Once the total bit count reaches to n, this module sets the *startAddMul* signal high and the n-bit input is fed to the addition (or multiplication) block. This enables the actual addition module to start addition (or multiplication for multiplication operation). Figure 6.8 shows the top-level views of *RAM_Fetch module*.

- **Addition module**: Once the *startAddMul* is set high, the addition module starts n-bit addition. The addition module consists of two 64-bit LUT[1] based carry-lookahead adders, cascaded together as a ripple carry adder. Addition is performed with this 128-bit ripple carry adder. Comparatively faster clock is used to fetch initial 64-bit and the first adder starts addition. Gradually, the next 64-bit is fetched in parallel and *startAdd2* is set high. Total 128-bit addition is performed taking consideration the carry-bit of the first adder. Finally, the addition result is stored in the encoded format to the output RAM. To avoid the timing overhead for RAM_Fetch

---

[1]The adder is realized inside the Look Up Tables (LUTs) of the FPGA.

operation. Fetch and addition operations are scheduled in parallel, hence the only overhead for RAM_Fetch module comes into account while starting the addition. Similar techniques can be applied while performing algorithms suitable for large bit multiplications (for example FFT multiplication).

Discussion on Implementation Results

m-bit ciphertext of k-subsequences with same bits and with each subsequences of length l_i are considered, where $l_i = n_i - 1$ if the subsequence consists of n_i same bits. Following Sect. 6.8.2 such subsequences can be stored by ($\lceil log n_i \rceil + 1$) bits due to encoding. However, if $l_i = 0$, then atleast 2 bits are required to store the encoded form (for sequences of $l_i = 1$ and $l_i = 2$, 2 and 3 bits are required). Hence, in general no saving is possible till $l_i > 3$ and the total saving of bits can be calculated as:

$$Total\ Saving = m - \sum_{i=0}^{k-1} (\lceil log n_i \rceil + 1) \qquad (6.4)$$

The saving of bits can be guaranteed if among k subsequences, atleast one subsequence is of length greater than 3. The probability of this event is quite high considering very large ciphertexts. While handling random ciphertexts, different subsequences can be of different lengths. Hence, while realizing in hardware, a concern is to decide the bit-size to store the cnt value in RAM_A and RAM_B, which stores the value length (l_i) for subsequences of repetitive bits.

6.8.3 Different Choices of Subsequence Length to Store in RAM

Figure 6.9 shows a relation between choice of pre-assumed length of subsequences with repetitive bits (l_i) and required RAM size to store the value. Due to the proposed encoding technique, the storage for encoded ciphertext are predecided based on l_i.

Choice of very large l_i may save storage, but randomly generated ciphertexts may consist of most of the subsequences with length l, where $l \neq l_i$. Hence, there may be a possible truncation (and introduction of error) due to lack of storage. Two cases may occur:

1. If subsequence length l in any case be such that $l > l_i$, then the subsequence can be divided in multiple RAM blocks.
2. If the ciphertext consists of multiple subsequences of length l such that $l < l_i$, that implies it requires larger RAM size (more number of RAM blocks) than dedicated. Hence, possible truncation of ciphertext and introduction of error is possible. To handle this error, it is checked if the maximum ciphertext bit length m has been reached while storing the ciphertext in RAM after encoding. For intermediate operations, if the dedicated RAM is full and all the m bits of the

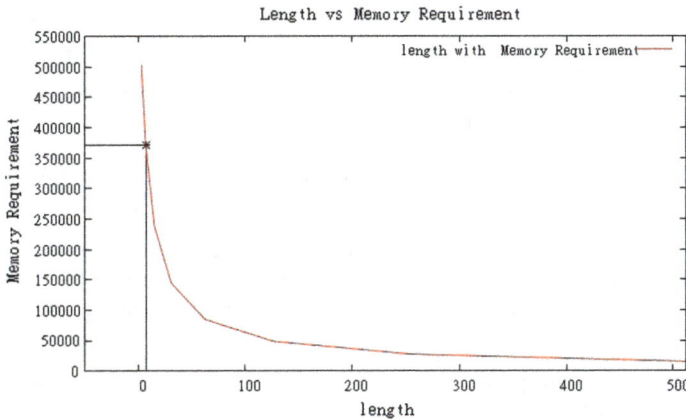

Fig. 6.9 Pre-assumed subsequence length versus required memory

ciphertext are still not stored in the RAM, then some of the RAM inputs are first
fetched for next operations and reuse those locations. If this oversize occurs in
the final result, then some more external storage support is required.

3. So far, we have only evaluated the encoding applied Addition module. Similar
   RAM Fetch module can be applied to implement multiplication, which we leave
   as a future area of research.

6.9 Conclusion

In this chapter, we discuss an encrypted URISC architecture with FHE as underlying
encryption scheme, which combines the flexibility of performing arbitrary operations
on encrypted data due to the property of FHE with design simplicity of URISC
architecture. Due to the use of FHE, randomization in memory handling and PC
branching solves the CPA Vulnerability issues of previous partial homomorphic
encryption based processor HEROIC. However in present literature, main challenge
for implementing homomorphic encryption schemes in actual hardware is to handle
huge parameter sets. Due to the large size of ciphertexts and evaluation keys, on-
chip storage of all data is impossible and external memory is mandatory. Memory
handling turns worse while working on reconfigurable platform and design of FHE
based processor is almost infeasible till date. At the end, an encoding scheme is
mentioned to store such large parameters in a compressed form, which requires less
storage. That may lead to improve the feasibility of actual FPGA realization of FHE
based ultimate encrypted processor in future.

References

Becker GT, Regazzoni F, Paar C, Burleson WP (2013) Stealthy dopant-level hardware trojans. In: Proceedings of the 15th international conference on cryptographic hardware and embedded systems, CHES'13. Springer, Berlin, pp 197–214

Bos JW, Lauter KE, Loftus J, Naehrig M (2013) Improved security for a ring-based fully homomorphic encryption scheme. In: Stam M (ed) IMA international conference, vol 8308. Lecture notes in computer science. Springer, Berlin, pp 45–64

Brenner M, Perl H, Smith M (2012a) How practical is homomorphically encrypted program execution? an implementation and performance evaluation. In: Proceedings of the 11th IEEE international conference on trust, security and privacy in computing and communications, TrustCom 2012, Liverpool, United Kingdom, 25–27 June 2012, pp 375–382

Brenner M, Perl H, Smith M (2012b) Practical applications of homomorphic encryption. In: SECRYPT 2012 - proceedings of the international conference on security and cryptography, Rome, Italy, 24–27 July 2012, pp 5–14

Breuer PT, Bowen JP (2014) Idea: towards a working fully homomorphic crypto-processor - practice and the secret computer. In: ESSoS, pp 131–140

Bruce Cousins David, Kurt Rohloff, Daniel Sumorok (2017) Designing an FPGA-accelerated homomorphic encryption co-processor. IEEE Trans Emerg Top Comput 5(2):193–206

Cao X, Moore C, O'Neill M, O'Sullivan E, Hanley N (2013) Accelerating fully homomorphic encryption over the integers with super-size hardware multiplier and modular reduction. IACR Cryptology ePrint Archive

Doröz Y, Öztürk E, Sunar B (2015) Accelerating fully homomorphic encryption in hardware. IEEE Trans Comput 64(6):1509–1521

Doröz Y, Öztürk E, Sunar B (2013) Evaluating the hardware performance of a million-bit multiplier. In: DSD. IEEE Computer Society, pp 955–962

Fletcher CW, Dijk Mv, Devadas S (2012) A secure processor architecture for encrypted computation on untrusted programs. In: Proceedings of the seventh ACM workshop on scalable trusted computing, STC '12, ACM, New York, NY, USA, pp 3–8

Gentry C, Halevi S (2011a) Fully homomorphic encryption without squashing using depth-3 arithmetic circuits. In: IEEE 52nd annual symposium on foundations of computer science, FOCS 2011, Palm Springs, CA, USA, 22–25 Oct 2011, pp 107–109

Gilreath WF, Laplante PA (2012) Computer architecture: a minimalist perspective. Springer Publishing Company Incorporated, Berlin

Goldwasser S, Kalai YT, Popa RA, Vaikuntanathan V, Zeldovich N (2013b) How to run turing machines on encrypted data. In: Advances in cryptology - CRYPTO 2013 - 33rd annual cryptology conference, Santa Barbara, CA, USA, 18–22 Aug 2013. Proceedings, Part II, pp 536–553

Hauck S, Wilson WD (1999) Runlength compression techniques for FPGA configurations. In: Proceedings of the 7th IEEE symposium on field-programmable custom computing machines (FCCM '99), pp 286–287

Irena Florencia, Murphy Daniel, Parameswaran Sri (2018) CryptoBlaze: a partially homomorphic processor with multiple instructions and non-deterministic encryption support. ASP-DAC 2018:702–708

Katz J, Lindell Y (2007) Introduction to modern cryptography (Chapman & Hall/CRC Cryptography and network security series). Chapman & Hall/CRC

Koblitz N, Menezes A, Vanstone S (2000) The state of elliptic curve cryptography. Des Codes Cryptogr 19(2–3):173–193

Library libScarab (2011). https://github.com/hcrypt-project/libscarab

Lindner R, Peikert C (2011) Better key sizes (and attacks) for lwe-based encryption. In: Proceedings of the 11th international conference on topics in cryptology: CT-RSA 2011, CT-RSA'11. Springer, Berlin, pp 319–339

Lyubashevsky V (2008) Lattice-Based identification schemes secure under active attacks, pp 162–179

Lyubashevsky V (2012) Lattice signatures without trapdoors. In: Proceedings of the 31st annual international conference on theory and applications of cryptographic techniques, EURO-CRYPT'12, Springer, Berlin, pp 738–755

Lyubashevsky V, Peikert C, Regev O (2010) On ideal lattices and learning with errors over rings. In: Proceedings of EUROCRYPT, volume 6110 of LNCS. Springer, pp 1–23

Nelson M (1991) The data compression book. Henry Holt and Co. Inc, New York

Pöppelmann T, Naehrig M, Putnam A, Macias A (2015) Accelerating homomorphic evaluation on reconfigurable hardware. Springer, Berlin, pp 143–163

Pöppelmann T, Naehrig M, Putnam A, Macías A (201) Accelerating homomorphic evaluation on reconfigurable hardware. In: Proceedings of the cryptographic hardware and embedded systems - CHES 2015, pp 143–163

Rivest RL, Shamir A, Adleman L (1978b) A method for obtaining digital signatures and public-key cryptosystems. Commun ACM 21(2):120–126

Shor PW (1994) Algorithms for quantum computation: discrete logarithms and factoring. In: Proceedings of the 35th annual symposium on foundations of computer science, SFCS '94. IEEE Computer Society, Washington, DC, USA, pp 124–134

Sinha Roy Sujoy, Frederik Vercauteren, Jo Vliegen, Verbauwhede, (2017) Ingrid hardware assisted fully homomorphic function evaluation and encrypted search. IEEE Trans Comput 66(9):1562–1572

Sinha Roy S, Järvinen K, Vercauteren F, Dimitrov V, Verbauwhede I (2015) Modular hardware architecture for somewhat homomorphic function evaluation. Springer, Berlin, pp 164–184

Tsoutsos NG, Maniatakos M (2013a) Investigating the application of one instruction set computing for encrypted data computation. In: SPACE, pp 21–37

Tsoutsos NG, Maniatakos M (2013b) Investigating the application of one instruction set computing for encrypted data computation. In: SPACE, pp 21–37

Tsoutsos NG, Maniatakos M (2014) Heroic: homomorphically encrypted one instruction computer. In: Proceedings of the conference on design, automation & test in Europe, pp 246:1–246:6

Wang W, Hu Y, Chen L, Huang X, Sunar B (2015) Exploring the feasibility of fully homomorphic encryption. IEEE Trans Comput 64(3):698–706

Wang W, Huang X (2013) Fpga implementation of a large-number multiplier for fully homomorphic encryption. In: ISCAS. IEEE, pp 2589–2592

Wang W, Hu Y, Chen L, Huang X, Sunar B (2012) Accelerating fully homomorphic encryption using gpu. In: HPEC. IEEE, pp 1–5

Chapter 7
Conclusion and Future Work

In this book, we have analyzed the prospect of realizing practical algorithms while working on FHE data. In this regard, we briefly reiterate our main contributions:

- In Chap. 3, feasibility of implementing different sorting techniques has been analyzed on encrypted data and a relation is established between the problem of sorting to security. Further, a two stage sorting technique specific to FHE data has been proposed to enhance the performance of encrypted sorting technique.
- Chapter 4 presents the method of translating traditional algorithms to encrypted domain and highlight challenges to implement such algorithms on existing unencrypted processor. Further, it shows how to handle different operators, abstract data types as well as loop handling, recursion problems while working with FHE data. However, detecting termination of such encrypted algorithm raises a pertinent question.
- Chapter 5 highlights aspects of secure database design with the power of homomorphic computation. This chapter mentions different incidents of IT-hacking on outsourced data in cloud and how different big names in IT industry have been affected due to such issues. Further, we discuss how to design basic and advanced homomorphic database operators and what are the performance challenges.
- In Chap. 6, design of an encrypted processors have been discussed. Mostly FHE based processor *FURISC* has been elaborated for outsourcing computations and achieving privacy. In this context, drawbacks of existing encrypted processors have been highlighted and it is explained how FHE ensures CPA security, while URISC architecture ensures advantages over multi-opcode FHE based processor. Finally, it is unveiled how design of different (partial, pseudo and FHE based) encrypted processors are required for different purposes. Somewhere security is compromised to achieve performance (in case of PHE based HEROIC), again for some other cases performance is a genuine bottleneck to support exhaustive computation.

© Springer Nature Singapore Pte Ltd. 2019
A. Chatterjee and K. M. M. Aung, *Fully Homomorphic Encryption
in Real World Applications*, Computer Architecture and Design Methodologies,
https://doi.org/10.1007/978-981-13-6393-1_7

Possible Future Works from Discussed Chapters

With all the previous discussions, few future works can be stemmed out:

- In Chap. 3, it is discussed how lazy sort is a promising technique in context of FHE sorting. However, all the sorting techniques discussed are effective for serial processors. As a future work, detailed implementation of sorting techniques suitable for parallel processors like bitonic sort, odd-even sort can be investigated. Such algorithms require huge number of concurrent processors or dedicated suitable hardware implementation to achieve best case performance. Hence, distinguished implementation approach should be acquired to realize such sorting techniques.
- As a first attempt in ciphertext size compression for FHE hardware module design, simple run-length encoding has been used, since it is easy to comprehend and implementation overhead is less. As a future extension. some complex encoding techniques can be considered for compression and possible overheads can be analyzed. Different variations of run-length encoding (like Lempel-Ziv, run-length encoding with re-ordering (Hauck and Wilson 1999)) can also be considered as a future encoding scheme.
- Using the proposed encoded building blocks an effort can be made to realize an overall FHE based cryptoprocessor in existing reconfigurable platform, which is still infeasible. Not only in case of FHE based implementations, usage of such compression technique can further be explored in case of implementing lattice based signature schemes with large key size (Güneysu et al. 2015).
- Fully homomorphic encryption scheme is prohibitively slow for most systems. In spite of recent developments on performance improvement, it is currently orders of magnitude slower than unencrypted computation. For this reason, other encryption schemes (for example: functional encryption) are till now preferred choice for cryptographers to perform arbitrary operations on encrypted data.
- FHE is important in different fields like national data security, healthcare and medical data security, banking and Internet of Things (IoT) data security. One important future work can be to design application specific homomorphic modules to provide domain specific optimizations for improved performance.

References

Güneysu T, Lyubashevsky V, Pöppelmann T (2015) Lattice-based signatures: optimization and implementation on reconfigurable hardware. IEEE Trans Comput 64(7):1954–1967

Hauck S, Wilson WD (1999) Runlength compression techniques for FPGA configurations. In: Proceedings of the 7th IEEE symposium on field-programmable custom computing machines (FCCM '99), pp 286–287

Appendix A
Lattice Based Cryptography

Lattice ($\mathscr{L}(\mathscr{B})$) is a discrete additive subgroup of $\mathbb{R}^m$, which corresponds to the vector space generated by all linear combinations with integer coefficients of a set $\mathscr{B} = \{\vec{b_0} \ldots \vec{b_{n-1}}\}$, with $b_i \in R^m$. It can be represented as:

$$\mathscr{L}(\mathscr{B}) = \sum_{i=0}^{n-1} z_i \vec{b_0} : z_i \in \mathbb{Z} \tag{A.1}$$

Basic $\mathscr{B}$ can be represented as matrix B with vectors $\vec{b_i}$ as rows and Eq. A.2 can be represented as:

$$\mathscr{L}(\mathscr{B}) = \sum_{i=0}^{n-1} \vec{z} \times B : \vec{z_i} \in \mathbb{Z}^n \tag{A.2}$$

Few key features of lattice are:

- Lattices can have an infinite number of bases for $n \geq 2$.
- If two matrices B_1 and B_2 are associated with same lattice, they are related by an integer matrix U such that $|det(U)| = 1$ Hence, absolute value of determinants are same for all the bases of any lattice and denoted as $det(\mathscr{L})$.
- Every lattice base B has corresponding half open parallelepiped $\mathscr{P}(B) \leftarrow \sum_{i=0}^{n-1} z_i \vec{b_i} : z_i \in (\frac{-1}{2}, \frac{1}{2}]\}$
- Two vectors in a lattice are congruent ($\vec{x} = \vec{y} \pmod{\mathscr{L}}$) if their difference is in the lattice ($\vec{x} - \vec{y}) \in \mathscr{L}$.
- The reduction of a vector y modulo a lattice base B, i.e ($\vec{x} = \vec{y} \pmod{\mathscr{B}}$) corresponds to determining $\vec{x} \in \mathscr{P}(B)$, that can be computed as: $\vec{x} = \vec{y} - \lceil \vec{y} \times B^{-1} \rceil \times B = \lceil \vec{y} \times B^{-1} \rceil \times B$

To understand why lattice is suitable for cryptographic purpose, next we shall discuss few unique property and problems of lattice (Fig. A.1).

© Springer Nature Singapore Pte Ltd. 2019
A. Chatterjee and K. M. M. Aung, *Fully Homomorphic Encryption in Real World Applications*, Computer Architecture and Design Methodologies,
https://doi.org/10.1007/978-981-13-6393-1

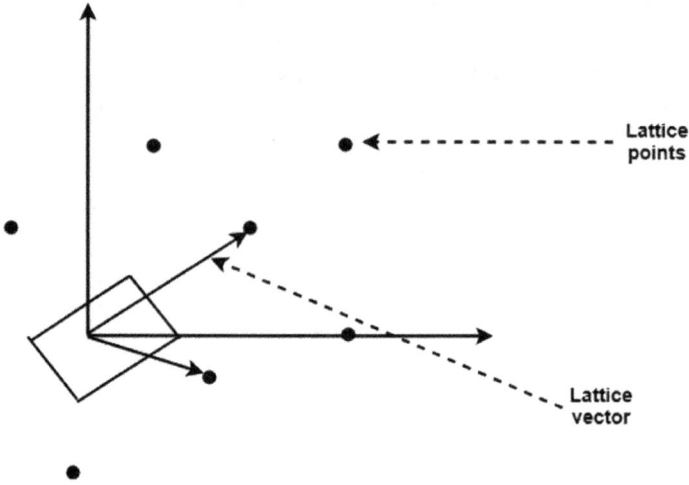

Fig. A.1 Lattice basics

Hermite Normal Form (HNF)

HNF base is unique to every lattice corresponds to a base H with following properties:

- $\forall i < j \ h_{i,j} = 0$
- $\forall j \ h_{j,j} > 0$
- $\forall i > j \ h_{i,j} \in (-h_{j,j}/2, h_{j,j}/2]$

HNF can be efficiently computed from any basis B of a lattice. Hence, this is considered a good choice for the public key of Lattice-Based Cryptosystems. Basic notion of security for Lattice-Based Cryptosystems are mostly dependent on the following:

- **Closest Vector Problem (CVP)**: Given a base $B \in \mathbb{R}^{n \times m}$ and $\overrightarrow{y} \in \mathbb{R}^m$, find $\overrightarrow{x} \in \mathscr{L}(B)$ such that $|| \overrightarrow{y} - \overrightarrow{x} || = min_{\overrightarrow{z} \in \mathscr{L}(B)} || \overrightarrow{y} - \overrightarrow{z} ||$.
- **Shortest Vector Problem (SVP)**: Given a base $B \in \mathbb{R}^{n \times m}$ a, find $\overrightarrow{x} \in \mathscr{L}(B)$ such that $|| \overrightarrow{x} || = min_{\overrightarrow{z} \in \mathscr{L}(B)} || \overrightarrow{z} ||$.
- **General Learning with Errors (GLWE)** (Brakerski et al. 2012): Let n, m, $q \in \mathbb{Z}$ and $R = \mathbb{Z}[t]/\phi_m(t)$, $R_q = R/qR$. Let χ be a Gaussian distribution over R. Given arbitrarily number of samples $(\overrightarrow{x_i}, y_i) \in R_q^{n+1}$, where $y_i = (\overrightarrow{x_i}, \overrightarrow{s}) + e_i$, where $\overrightarrow{x_i}, \overrightarrow{s} \leftarrow R_q^n$ sampled uniformly and $e_i \leftarrow \chi$, find $\overrightarrow{s}$.

Both LWE and lattices are connected in the following manner: Let us consider lattice $\mathscr{L}(B)$ where the matrix $B \in \mathbb{Z}^{n \times t}$ has t numbers of $\overrightarrow{x_i}$ samples as columns. If closest vector $\overrightarrow{y'}$ can be computed to y with these samples that is equivalent to have a solution to LWE problem.

Reference

Brakerski Z, Gentry C, Vaikuntanathan V (2012) (leveled) Fully homomorphic encryption without bootstrapping. In: Innovations in theoretical computer science, pp 309–325

Appendix B
LWE Based FHE

Initial proposed FHE schemes are based on Gentry's seminal work with strong computational assumptions. These schemes are simplified with the learning with errors (LWE) security assumption (Regev et al. 2006). In the subsequent sections, we first mention the basics of LWE based schemes and discuss few basic constructions of LWE based FHE.

B.1 Basics of LWE Based Cryptosystem

LWE was first introduced by Regev et al. (2006) generalizing learning parity with noise problem. For positive integers n and $q \geq 2$, vector $s \in \mathbb{Z}_n^q$, probability distribution χ on $\mathbb{Z}_{11}^{\mathbb{K}}$, let $A_{s,\chi}$ be the distribution obtained by choosing a random vector $a \leftarrow \mathbb{Z}_q^n$ uniformly and a noise term $e \leftarrow \chi$ and outputting $(a, [(a, s) + e]_q) \in \mathbb{Z}_q^n \times \mathbb{Z}_q$. The decision version of LWE (DLWE) is to distinguish between noisy inner products and uniformly random samples from $\mathbb{Z}_q^n \times \mathbb{Z}_q$. It is defined as follows:

Definition B.1 For an integer $q = q(n)$ and an error distribution $\chi = \chi(n)$ over $\mathbb{Z}$, $DLWE_{n,m,q,\chi}$ is a problem to distinguish with non-negligible advantage, m samples chosen according to $A_{s,\chi}$ from m samples chosen from uniform distribution over $\mathbb{Z}_q^n \times \mathbb{Z}_q$. In this variant, the adversary gets oracle access to $A_{s,\chi}$.

It is interesting to note how this LWE problem can form the basic notion of security of cryptosystem. In general, LWE based crypto schemes generate ciphertexts $\mathbb{Z}_q$ for modulus q, which upon decryption produces noisy version of the message. Noise is added at encryption for security purposes. The decryption process is essentially computing an inner product of the ciphertext and the secret key vector. The noise should be below a certain threshold to retain the homomorphic property of the scheme. As discussed in Regev et al. (2006), parameters of LWE are mapped to any cryptosystem in the following way:

© Springer Nature Singapore Pte Ltd. 2019

A. Chatterjee and K. M. M. Aung, *Fully Homomorphic Encryption in Real World Applications*, Computer Architecture and Design Methodologies, https://doi.org/10.1007/978-981-13-6393-1

- **Secret Key**: Private key s is chosen as $s \in \mathbb{Z}_q^n$.
- **Public Key**: For $i = 1 \ldots m$, m vectors are chosen at random $a_1, \ldots a_m \in \mathbb{Z}_q^n$ uniformly from the distribution. $e_1, \ldots e_m \in \mathbb{Z}_q$ independently according to χ. Public key is considered as $(a_i, m_i)_{i=1}^m$ where $b_i = (a_i, s) + e_i$.
- **Encryption**: To encrypt a bit, random set S is chosen uniformly among 2^m subsets of $[m]$. The ciphertext (c) is:

$$c = \left(\sum_{i \in S} a_i, \sum_{i \in S} b_i \right), \textit{ for bit} = 0 \tag{B.1}$$

$$c = \left(\sum_{i \in S} a_i, \lfloor \frac{p}{2} \rfloor + \sum_{i \in S} b_i \right), \textit{ for bit} = 1 \tag{B.2}$$

- **Decryption**: Decryption of the ciphertext term with (a, b) pair is closer to 0 than $\lfloor \frac{p}{2} \rfloor$ if $b - (a, s)$ is 0, else 1.

B.2 Important LWE Based FHE Schemes

One of the notable scheme solely based on LWE assumption is explained in Brakerski et al. (2011), which shows a new simplified noise handling technique called relinearization supporting SHE scheme with much simpler hardness assumptions than ideals. Another important feature of this scheme is the formation of FHE from this SHE scheme without costly squashing or assumption of subset sum problem as detailed in Brakerski et al. (2011).

In this scheme, ciphertext $\overrightarrow{c}$ and secret key $\overrightarrow{s}$ are n dimensional vectors, where dot vectors of $(\overrightarrow{c}, \overrightarrow{s}) \approx \mu$ with small error that is removed by rounding. In this process, multiplication blows up the ciphertext size. Relinearization is a procedure that takes the long ciphertext that encrypts $\mu_1.\mu_2$ under a long key $\overrightarrow{s} \otimes \overrightarrow{s}$. This further compresses into a normal-sized n-dimensional ciphertext under a normal-sized n-dimensional key $\overrightarrow{s}$. This scheme is further improved by Craig Gentry, Amit Sahai and Brent Waters in their notable contribution in the paper (Gentry et al. 2013b). Now onwards we term that as GSW scheme.

Basics of GSW

- In this scheme, the requirement of relinearization has been removed using matrix multiplication using sub-cubic computation.
- This scheme proposes an identity-based FHE scheme, in which user with only the public parameters should be able to perform both encryption and homomorphism operations. The homomorphism operations should allow a user to take two ciphertexts encrypted to the same target identity, and homomorphically combine them to produce another ciphertext under the same target identity.
- These scheme can further be extended for a construction of homomorphic attribute-based encryption (ABE) with minor modifications.

References

Brakerski Z, Vaikuntanathan V (2011) Efficient fully homomorphic encryption from (Standard) LWE. FOCS 97–106

Gentry C, Sahai A, Waters B (2013b) Homomorphic encryption from learning with errors: conceptually-simpler, asymptotically-faster. Attribute-based. CRYPTO 75–92

Regev O (2006) Lattice-based cryptography. CRYPTO 131–141

Appendix C
GSW Based FHE Approach

Series of research contributions have been made in the direction of Gentry proposed homomorphic scheme (as discussed in BGV scheme Brakerski et al. 2012 and others) based on lattice based assumptions. In this contributions, circular security is one of the important assumption to obtain a FHE scheme from a leveled HE scheme. However, this is the main bottleneck in terms of performance while applying FHE in real world applications.

Another GSW based direction of research includes achieving the power of fully homomorphic computation in the simplest setting for bit-wise computation. Ducas et. al in their work (Ducas and Micciancio 2015) has proposed such FHE scheme where bootstrapping can be computed in less than seconds in bit level. In Table C.1, a comparison has been shown between such BGV and GSW based schemes. BGV based FHE construction details are discussed in chapter 2. In the subsequence section, we provide a brief idea of FHE schemes where bit-wise homomorphy is supported.

Table C.1 Present FHE trends in literature

	BGV based (Brakerski et al. 2012)	GSW based (Gentry et al. 2013a)
Performance	Slow in operation but processes huge number of bits	Faster operation but single bit processing
Operations	Limited set of operations due to parameter set:	Can support any arbitrary operation
Limitations	Operations can not be done 1. Comparison 2. Bit Extraction related operations 3. Addition etc	

© Springer Nature Singapore Pte Ltd. 2019
A. Chatterjee and K. M. M. Aung, *Fully Homomorphic Encryption in Real World Applications*, Computer Architecture and Design Methodologies, https://doi.org/10.1007/978-981-13-6393-1

Table C.2 Modulo 2 to Modulo 4 translation

m_1	m_2	$Dec[E(m_1 + m_2)]$ (modulo 2)	$Dec[E(m_1 \overline{\wedge} m_2)]$ (modulo 4)
$E(0)$	$E(0)$	0	1
$E(0)$	$E(1)$	1	0
$E(1)$	$E(0)$	1	0
$E(1)$	$E(1)$	0	0

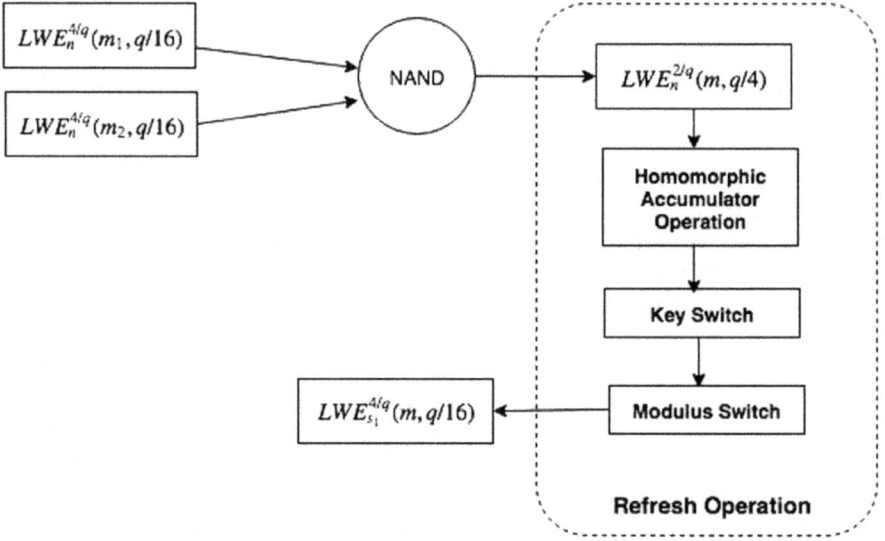

Fig. C.1 FHEW refreshing steps

C.1 Brief Details of Scheme Supporting Bit-Wise Homomorphy

Given two encrypted bits $E(b_1)$ and $E(b_2)$, this scheme aims to compute logical NAND of the two bits with following observation as shown in Table C.2. Computing $E(m_1 + m_2)$ in modulo 2 allows to homomorphically compute exclusive-or of two bits. From moving arithmetic modulo 2 to modulo 4, logical NAND computation is done during the bootstrapping process of this scheme. That indicates adding $E(m_1)$ and $E(m_2)$ generates $E(m)$, such that $E(m)$ of $m = 2$, if $m_1 \overline{\wedge} m_2 = 0$ or $m \in (0, 1)$, if $m_1 \overline{\wedge} m_2 = 1$.

This scheme homomorphically computes the NAND of two LWE encryptions. The noise introduced in this case is much lower, hence the refreshing technique is not so costly. The next step is the refreshing technique as shown in Fig. C.1. homomorphically evaluates $LWE_s^2(m, q/4)$ to $LWE_s^4(m, q/16)$. Details of the procedure can be found in the paper (Ducas and Micciancio 2015).

$$Refresh : LWE_s^2(m, q/4) \rightarrow LWE_s^4(m, q/16) \tag{C.1}$$

This scheme has been further enhanced in the paper (Chillotti et al. 2016a) where the bootstrapping has been improved from less than 1 s to less than 0.1 s. Some further improvements have been proposed in Chillotti et al. (2017a) with suitable packing techniques. These works form the mathematical background of TFHE library as discussed in the FHE library section.

References

Brakerski Z, Gentry C, Vaikuntanathan V (2012) (leveled) Fully homomorphic encryption without bootstrapping. In: Innovations in theoretical computer science, pp 309–325

Chillotti I, Gama N, Goubin L (2016a) Attacking FHE-based applications by software fault injections. IACR cryptology ePrint archive, vol 1164

Chillotti I, Gama N, Georgieva M, Izabachène M (2017a) Faster packed homomorphic operations and efficient circuit bootstrapping for TFHE. ASIACRYPT 377–408

Ducas L, Micciancio D (2015) FHEW: bootstrapping homomorphic encryption in less than a second. EUROCRYPT 617–640

Gentry C, Sahai A, Waters B (2013a) Homomorphic encryption from learning with errors: conceptually-simpler, asymptotically-faster, attribute-based. In: Advances in cryptology - CRYPTO 2013 - 33rd Annual cryptology conference, pp 75–92

Appendix D
FHE Based Libraries in Literature

Few FHE based libraries have been reported in literature with different features mentioned in Table D.1. In this section, we detail few of them:

Library HElib

HElib (Halevi et al. 2013b) is c++ based FHE software library based on Brakerski-Gentry-Vaikuntanathan (BGV) scheme (Brakerski et al. 2012) along with few optimizations. The library mostly focuses on the optimizations present in the paper (Gentry et al. 2012) and packing techniques from Smart and Vercauteren (2011). Work in Smart and Vercauteren (2011) presented a modified version of Gentry's fully homomorphic public key encryption scheme which supports SIMD style operations. This paper shows how to select parameters to enable such SIMD operations, whilst still maintaining practicality of the key generation technique of Gentry and Halevi. The proposed somewhat homomorphic scheme can be made fully homomorphic by recrypting all data elements seperately. However, this paper has shown a SIMD approach that can be used to perform recrypt in parallel supporting improved performance. this paper also demonstrates implementing AES homomorphically with this library.

Table D.1 FHE libraries in literature

Libraries	Scheme	Supporting libs
LibScarab (Perl et al. 2011)	SV (Smart and Vercauteren 2010)	GMP, FLINT MPFR, MPIR
HElib (Halevi et al. 2013b)	BGV (Brakerski et al. 2012)	NTL, GMP
FHEW (Ducas and Micciancio 2014)	FHEW (Ducas and Micciancio 2015)	FFTW
TFHE (Chillotti et al. 2017b)	TFHE	FFTW
SEAL (Laine et al. 2017)	FV12 (Fan and Vercauteren 2012)	No external dependency

© Springer Nature Singapore Pte Ltd. 2019
A. Chatterjee and K. M. M. Aung, *Fully Homomorphic Encryption in Real World Applications*, Computer Architecture and Design Methodologies,
https://doi.org/10.1007/978-981-13-6393-1

Library FHEW

FHEW is open-source FHE library mathematically based on the paper "FHE boot-strapping in less than a second" (Ducas and Micciancio 2015). The name FHEW that is "Fastest Homomorphic Encryption in the West" is more of a reference to FFTW ("Fastest Fourier Transform in the West") than a claim about performance. In the paper, authors proposed method to homomorphically compute simple bit operations, and refresh (bootstrap) the resulting output within just about half a second in a consumer grade personal computer.

Most interesting feature of this library is that it provides bootstrapping in simplest possible setting which proved to be benificial later on in case of designing complex encrypted algorithms. To provide homomorphy, this library follows simple steps of encrypting single bits, and evaluating boolean NAND circuits on them. Basic idea is as follows: Given two encrypted bits $E(m_1)$ and $E(m_1)$, computing noisier version $E(m_1 + m_2)$ is very straight-forward. For arithmetic modulo 2, this is equivalent to compute exclusive-or of two bits. Next, logical NAND computation can be done by extending modulo 2 to modulo 4 with a bootstrapping (refreshing) technique.

Library TFHE

TFHE or "Fast Fully Homomorphic Encryption over the Torus" is another open-source library for FHE broadly a modification of FHEW. The mathematical background of this work is detailed in the paper "Faster fully homomorphic encryption: Bootstrapping in less than 0.1 s" (Chillotti et al. 2017b). This library supports the homomorphic evaluation of the 10 binary gates along with negation and encrypted multiplexer gate. Each binary gate takes about 13 ms single-core time to evaluate and the multiplexer gate takes about 26 CPU-ms. The gate-bootstrapping of TFHE has no restriction on the number of gates or on their composition that gives flexibility to realize real world design in encrypted form.

Library cuHE

Library cuHE or CUDA Homomorphic Encryption Library (cuHE Library 2018) is a GPU-accelerated library for HE schemes and homomorphic algorithms defined over polynomial rings. This library shows different techniques of memory minimization, memory and thread scheduling and low level CUDA related optimizations to take full advantage of the mass parallelism and high memory bandwidth of GPUs. This library is mostly SWHE with level limitations of HE.

References

Brakerski Z, Gentry C, Vaikuntanathan V (2012) (leveled) Fully homomorphic encryption without bootstrapping. In: Innovations in theoretical computer science, pp 309–325

Chillotti I, Gama N, Georgieva M, Izabachène M (2017b) TFHE: fast fully homomorphic encryption library over the torus. Retrieved from https://github.com/tfhe/tfhe (accessed September 2017)

cuHE. https://github.com/vernamlab/cuHE, Last Accessed: 11.10.2018

Ducas L, Micciancio D (2014) A fully homomorphic encryption library. Retrieved from https://github.com/lducas/FHEW (accessed October 2018)

Ducas L, Micciancio D (2015) FHEW: bootstrapping homomorphic encryption in less than a second. EUROCRYPT 617–640

Fan J, Vercauteren F (2012) Somewhat practical fully homomorphic encryption. Cryptology ePrint archive, Report 2012/144. Retrieved from http://eprint.iacr.org/2012/144

Gentry C, Halevi S, Smart NP (2012) Better bootstrapping in fully homomorphic encryption. In: Public key cryptography - PKC 2012 - 15th International conference on practice and theory in public key cryptography, pp 1–16

Halevi S, Shoup V (2013b) An implementation of homomorphic encryption. Retrieved from https://github.com/shaih/HElib (accessed September 2018)

Laine K, Chen H, Player R (2017) Simple encrypted arithmetic library. Retrieved from https://sealcrypto.codeplex.com/ (accessed September 2018)

Perl H, Brenner M, Smith M (2011) Poster: an implementation of the fully homomorphic smart-vercauteren crypto-system. In: ACM conference on computer and communications security, pp 837–840

Smart NP, Vercauteren F (2010) Fully homomorphic encryption with relatively small key and ciphertext sizes. In: Proceedings of the 13th international conference on practice and theory in public key cryptography, PKC'10, pp 420–443

Smart NP, Vercauteren F (2011) Fully homomorphic SIMD operations. IACR cryptology ePrint archive

Appendix E
Attacks on SWHE and FHE

In spite of the fact that HE specially FHE is considered to be the holy grail of cryptography, it is important to note that FHE is also susceptible to different kinds of attacks. Following the discussion of Martins et al. (2018), here we mention few important attacks against SWHE and FHE. In the next sections, the attacks are largely classified in two types: Passive and Active attacks.

E.1 Passive Attacks Against HE

In this section, we start our discussion about attacks against some known SWHE schemes like RSA and El-Gamal.

- **Attacks against RSA**: The notion of RSA security lies on the hardness of factoring large integers. However, referring to RSA discussion in Chap. 2, an attacker can find the secret key d by factoring modulus q since e is known. In the work (Boneh et al. 1999), brute-force attack on RSA has been shown to obtain the secret key. Most known and efficient algorithm about integer factorization is Pollard's General Number Field Sieve (GNFS) (Lenstra et al. 1990). Table E.1 mentions few notable contributions in this direction.
- **Attack against El-gamal cryptosystem**: Compared to integer factorization problem, large amount of research has been done in the direction of solving discrete logarithm problem which is the main notion of security for El-gamal Cryptosystem. Few notable contributions in this direction are highlighted in Joux et al. (2014), Menezes et al. (1996).
- **Attack against Lattice based cryptosystem (LBC)**: Lattice based cryptography is mostly dependent on two mathematical problems CVP and SVP. Overview of CVP and SVP are given in previous discussions. Few Attacks against LBC are mentioned in Table E.2 mostly against SVP and CVP problems.

© Springer Nature Singapore Pte Ltd. 2019

A. Chatterjee and K. M. M. Aung, *Fully Homomorphic Encryption in Real World Applications*, Computer Architecture and Design Methodologies, https://doi.org/10.1007/978-981-13-6393-1

Table E.1 Different passive attacks

Algorithm	Attack
Integer factorization	Pollard's general number field sieve (Lenstra et al. 1990)
Discrete logarithm problem	Shor algorithm and others (Joux et al. 2014;Shor 1990)
Lattice based attacks	
NTRU and variant of Gentry's scheme	Exploitation of ring structure leads to subexponential and quantum polynomial attacks (Albrecht et al. 2016; Cramer et al. 2016)
SVP	SVP solvers (Kuo et al. 2016)
CVP, SVP, RSA, AGCD problem	Lattice basis reduction algorithms (Lenstra et al. 1982)

Table E.2 Few active attacks against HE

Scheme	Attacks
Schemes with either HE additions or multiplications	IND-CCA1 secured but not IND-CCA2 (Lipmaa et al. 2008)
AGCD-, LWE-, and NTRU-based schemes	key recovery attacks [adversary is capable of getting the secret key with a polynomial amount of queries to a decryption oracle]. (Dahab et al. 2015)
Variant of SV SHE scheme in	IND-CCA1 secured but not IND-CCA2. (Loftus et al. 2010)

E.2 Active Attacks Against HE

All the attacks mentioned in the previous section refer to passive attacks which does not require any direct interference of the attacker with the target system. On the other hand, active attacks are based on the simple assumption that: If an attacker can identify which of two possible plaintexts between m_0 and m_1 can encrypt c with a probability more than 0.5, then the system is considered to be insecure. This type of security requirements can be conceptualized with few indistinguishably assumptions like:

- **Chosen plaintext attack** or **IND-CPA**
- **chosen ciphertext attack** or **IND-CCA1**
- **adaptive chosen ciphertext attack** or **IND-CCA2**

Semantic security assumptions are nicely explained in Loftus et al. (2011) by defining game between a challenger and an adversary A. For FHE, an attacker can decrypt arbitrary ciphertexts and the secret key is made public in an encrypted form.

Hence, FHE is not considered as IND-CCA1 and IND-CCA2 secured but IND-CPA secured. However, some of the SWHE schemes has proved to be IND-CCA1 secured (Loftus et al. 2011). Finally, in the paper (Chillotti et al. 2016a) authors have shown how FHE based applications can be susceptible to software fault injection attacks due to its ability to compute function.

References

Albrecht M, Bai S, Ducas L (2016) A subfield lattice attack on overstretched NTRU assumptions. Springer, Berlin, pp 153–178

Boneh D (1999) Twenty years of attacks on the RSA cryptosystem. Not AMS 46:203–213

Chillotti I, Gama N, Goubin L (2016a) Attacking FHE-based applications by software fault injections. IACR cryptology ePrint archive, vol 1164

Cramer R, Ducas L, Peikert C, Regev O (2016) Recovering short generators of principal ideals in cyclotomic rings. Springer, Berlin, pp 559–585

Dahab R, Galbraith S, Morais E (2015) Adaptive key recovery attacks on NTRU-based somewhat homomorphic encryption schemes. Cryptology ePrint archive, Report 2015/127. Retrieved from http://eprint.iacr.org/

Joux A, Odlyzko A, Pierrot C (2014) The past, evolving present, and future of the discrete logarithm. In: Open problems in mathematics and computational science. Springer International Publishing, Cham, pp 5–36

Kuo P-C, Schneider M, Dagdelen Ö, Reichelt J, Buchmann J, Cheng C-M, Yang B-Y (2011) Extreme enumeration on GPU and in clouds: how many dollars you need to break SVP challenges. In: Proceedings of the 13th international conference on cryptographic hardware and embedded systems (CHES)

Lenstra AK, Lenstra HW Jr, Lovsz L (1982) Factoring polynomials with rational coefficients. Math Ann 261:515–534

Lenstra AK, Lenstra HW Jr, Manasse MS, Pollard JM (1990) The number field sieve. In: STOC, pp 564–572

Lipmaa H (2008) On the CCA1-security of elgamal and Damgård's elgamal. IACR cryptology ePrint archive, vol 234

Loftus J, May A, Smart NP, Vercauteren F (2010) On CCA-secure fully homomorphic encryption. Cryptology ePrint archive, Report 2010/560. Retrieved from http://eprint.iacr.org/

Loftus J, May A, Smart NP, Vercauteren F (2011) On CCA-secure somewhat homomorphic encryption. Selected areas in cryptography, pp 55–72

Martins P, Sousa L, Mariano A (2018) A survey on fully homomorphic encryption: an engineering perspective. ACM Comput Surv 50(6):83:1–83:33

Menezes AJ, Vanstone SA, Van Oorschot PC (1996) Handbook of applied cryptography. CRC Press, Boca Raton

Shor PW (1994) Algorithms for quantum computation: discrete logarithms and factoring. In: Proceedings of the 35th annual symposium on foundations of computer science, pp 124–134 (1990)

Appendix F
Examples of Homomorphic Real World Applications

In this final section, we mention (in Table F.1 and in Table F.2) few real world problems where power of homorphic encrypted computation is extensively used. Design of these encrypted applications rightly justify the motivation of our book which discusses the steps and challenges of realizing traditional algorithms to their encrypted counterpart.

Table F.1 FHE real world applications

Domain	Application
Medical	Encrypted cardiac risk factor algorithm (Carpov et al. 2016)
Medical	Encrypted predictive analysis on medical data (Bos et al. 2014)
Medical	Long-term patient monitoring via cloud-based ECG data acquisition and encrypted analytics design (Kocabas et al. 2014)
Analytics [useful for business and medical]	Logistic regression over encrypted data (Chen et al. 2018)
Analytics	Big data analytics over encrypted datasets with seabed (Papadimitriou et al. 2016)
Analytics	Encrypted classification in machine learning (Graepel et al. 2012)
Data analysis	Encrypted statistical analysis (Lu et al. 2012)
Deep learning	Running convolutional neural networks (CNN) on encrypted data (Badawi et al. 2018)

© Springer Nature Singapore Pte Ltd. 2019

A. Chatterjee and K. M. M. Aung, *Fully Homomorphic Encryption in Real World Applications*, Computer Architecture and Design Methodologies, https://doi.org/10.1007/978-981-13-6393-1

Table F.2 FHE real world applications

Domain	Application
Financial	Encrypted financial computational model for cloud framework (Peng et al. 2016)
Cyber Physical System (CPS)	Primitives for computations on encrypted data for CPS systems (Hu et al. 2016)
Cyber Physical System (CPS)	Encrypting controller using FHE for security of cyber-physical systems (Kim et al. 2016)
Others	Secure distributed incremental information aggregation for smart grids using HE (Alabdulatif et al. 2017)
Others	Secure friend discovery in social strength-aware Proximity-Based Mobile Social Networks (PMSNs) (Niu et al. 2015)

References

Alabdulatif A, Kumarage H, Khalil I, Atiquzzaman M, Yi X (2017) Privacy-preserving cloud-based billing with lightweight homomorphic encryption for sensor-enabled smart grid infrastructure. IET Wirel Sens Syst 7(6):182–190

Badawi AA, Chao J, Lin J, Mun CF, Jie SJ, Tan BHM, Nan X, Aung KMM, Chandrasekhar VR (2018) The AlexNet moment for homomorphic encryption: HCNN, the first homomorphic CNN on encrypted data with GPUs. IACR cryptology ePrint archive, vol 1056

Bos JW, Lauter KE, Naehrig M (2014) Private predictive analysis on encrypted medical data. J Biomed Inform 50:234–243

Carpov S, Nguyen TH, Sirdey R, Costantino G, Martinelli F (2016) Practical privacy-preserving medical diagnosis using homomorphic encryption. CLOUD 593–599

Chen H, Gilad-Bachrach R, Han K, Huang Z, Jalali A, Laine K, Lauter KE (2018) Logistic regression over encrypted data from fully homomorphic encryption. IACR cryptology ePrint archive, vol 462

Graepel T, Lauter KE, Naehrig M (2012) ML confidential: machine learning on encrypted data. ICISC 1–21

Hu P, Mukherjee T, Valliappan A, Radziszowski S (2016) Evaluation of homomorphic primitives for computations on encrypted data for CPS systems. In: Smart city security and privacy workshop (SCSP-W), pp 1–5

Kim J, Lee C, Shim H, Cheon JH, Kim A, Kim M, Song Y (2016) Encrypting controller using fully homomorphic encryption for security of cyber-physical systems. IFAC-PapersOnLine 49(22):175–180

Kocabas O, Soyata T (2014) Private predictive analysis on encrypted medical data. J Biomed Inform 50:234–243. https://doi.org/10.4018/978-1-4666-5864-6.ch019

Lu W, Kawasaki S, Sakuma J (2017) Using fully homomorphic encryption for statistical analysis of categorical, ordinal and numerical data, NDSS

Niu B, He Y, Li F, Li H (2015) Achieving secure friend discovery in social strength-aware PMSNs. MILCOM 947–953

Papadimitriou A, Bhagwan R, Chandran N, Ramjee R, Haeberlen A, Singh H, Modi A, Badrinarayanan S (2016) Big data analytics over encrypted datasets with seabed. OSDI 587–602

Peng H-T, Hsu WWY, Ho J-M, Yu M-R (2016) Homomorphic encryption application on FinancialCloud framework. SSCI 1–5

Bibliography

Chillotti I, Gama N, Georgieva M, Izabachène M (2016b) Faster fully homomorphic encryption: bootstrapping in less than 0.1 seconds. In: ASIACRYPT (1), pp 3–33
UniCrypt. https://github.com/bfhevg/unicrypt/blob/master/README.md, Last Accessed: 11.10.2018

CPSIA information can be obtained
at www.ICGtesting.com
Printed in the USA
LVHW080827040419
612950LV00001B/53/P

9 789811 363924